# BIORESOURCES
## Conservation Strategies

# BIORESOURCES
## Conservation Strategies

*Editors*
**V. Sashi**
**S. Poornima**



**Narosa Publishing House**

New Delhi Chennai Mumbai Kolkata

**Bioresources**
Conservation Strategies
110 pgs.

*Editors*
**V. Sashi**
Associate Professor
**S. Poornima**
Assistant Professor

Department of Botany
PSGR Krishnammal College for Women
Coimbatore



NAROSA PUBLISHING HOUSE PVT. LTD.

22 Delhi Medical Association Road, Daryaganj, New Delhi 110 002
35-36 Greams Road, Thousand Lights, Chennai 600 006
306 Shiv Centre, Sector 17, Vashi, Navi Mumbai 400 703
2F-2G Shivam Chambers, 53 Syed Amir Ali Avenue, Kolkata 700 019

**www.narosa.com**



Printed from the camera-ready copy provided by the Editors.

ISBN 978-81-8487-363-4

Published by N.K. Mehra for Narosa Publishing House Pvt. Ltd.,
22 Delhi Medical Association Road, Daryaganj, New Delhi 110 002

Printed in India

# Preface

To commomerate the Golden Jubilee celebrations an International Conference on Conservation Strategies for Bioresources was organized by the Departments of Botany (DST – FIST sponsored) and Zoology, PSGR Krishnammal College for Women. The conference was sponsored by UGC, DBT, CSIR and DRDO. The conference at international level has been aimed to serve as a platform to present the innovative research carried out across the globe and to create awareness on the values of bioresources and the need for their conservation. The thrust areas are exploration of bioresources for food, medicine, fiber and biofuels, conservation strategies through biotechnology- *in vitro, in vivo,* marker assisted breeding, IPR and patents. There was an over whelming response from researchers and academicians around the world.

I am deeply indebted to our Managing Trusty Sri. G. Rangaswamy, Secretary Smt. R. Nandini, Principal Dr. (Mrs.) N. Yesodha Devi and all my colleagues of PSGR Krishnammal College for Women for their guidance, encouragement and co-operation in bringing out this proceedings. I whole heartedly thank all the members of the International and National advisory board for their timely help in organizing the conference in an effective manner.

I thank immensely Mrs. S. Vasandha, Dean, Student affairs and Former Head, Department of Botany, PSGR Krishnammal College for Women for her constant support and guidance. My heartfelt thanks goes to the authorities of all the funding agencies for their financial support. Finally I thank Narosa Publishing House Pvt. Ltd., for the excellent work in the publication of the conference proceedings.

**V. Sashi**

# Contents

*Preface* *v*

**Biodiversity (Bio-Resource) Conservation: A Global Initiative** **1**
*Oommen V. Oommen*

**Bioprospecting bioresources: Issues related to environmental sustainability** **4**
*M. N. V. Prasad*

**Conservation of genetic resources and identifying genotypes using molecular markers in *Prunus*: Recent advances in peach and plum** **6**
*Jayasankar S., El-Sharkawy I., Sherif S., El-sadr H.* and *Dinesh Kumar S.*

**Genetic analysis for conservation and improvement of important wood traits in *Dalbergia sissoo roxb.*** **7**
*Ashok Kumar*

**Mechanism of salt exclusion and secretion in the mangrove *Avicennia*** **14**
*Pannaga Krishnamurthy, Pavithra AJ, Wee-Kee Tan, Teck-Kwang Lim, Tit-Meng Lim, Qingsong Lin, Chiang-Shiong Loh and Prakash P. Kumar*

**Conservation strategies for plant genetic resources with special reference to agro-biodiversity in India** **16**
*N. Sivaraj*

**Orchids in the adjoining areas of Shimla, Himachal Pradesh: Distribution, survival threats and some conservation measures** **23**
*Ravi Kant and Ashli R.*

**Conservation of indigenous medicinal mushroom *Ganoderma lucidum*: An approach** **27**
*Uma Gowrie S., G. Chathurdevi* and *K. Rani*

**Effect of Jasmonic acid and Salicylic acid on callus initiation and organogenesis in *Vigna mungo* (L.) Hepper using hypocotyl explant** **41**
*Asha A., S. P. Vairamathi* and *K. Lingakumar*

**Effect of wild plant extracts against *Fusarium solani* and Alternaria alternata in *in vitro*.** 49
*R. R. Saswade*

**Cellulase and xylanase enzyme activities in the basidiocarp of *Pleurotus spp*** 54
*Padmaja C. K.* and *Shiji Wilson*

**Pathogenicity of root-knot nematode, *Meloidogyne incognita* on chilli plant (CV.NS-230)** 59
*Bhavani S.* and *Shyamala Vijayavathi B.*

**Preliminary phytochemical and biochemical analysis of *Bacopa monniera*** 62
*Gopika S. and K. Hemalatha*

**Effects of stressors on primary and secondary stress responses in *Cirrhinus mrigala* and its management using a herbal stress reliever stresroak** 67
*Vasanthi J. and S. Binukumari*

**Antibacterial activity of *Centella asiatica* (L.)** 70
*T. V. Binu* and *C. D. Varghese*

**Studies on phytoremediation using *Vetiveria zizanoides* and *Eichhornia crassipes* on sewage water** 73
*C. R. Hima, M. Manimegalai* and *K. Shanthi*

***In-vivo* analgesic activity of methanolic leaf extract of *Molineria recurvata* on albino rat** 78
*Prasanta Dey, Manidipa Mukherjee, Ranjib Ghosh, Rajkumar Majumder* and *Tejendra Bhakta*

**Conservation and management aspects of the sacred groves in Mahe, U.T. of Puducherry** 81
*K. Sasikala, G. Pradeepkumar, C. C. Harilal* and *C. P. Ravindran*

**Mosquito larvicidal activity of *Momordica charantia* leaf extract against filarial vector *Culex quinquesfasciatus*** 86
*Annapoorani C. A.* and *Manimegalai K.*

**Studies on larvicidal activity of *Gloriosa superba* on the development stages of the Chikungunia mosquito, Aedes aegypti** 90
*C. R. Hima* and *M. Manimegalai*

**Effect of green manure and biofertilizer on the biochemcial parameters of cluster beans (*Cyamopsis tetragonoloba* L. var. PNB)** 94
*M. V. Anitha, B. Vijayakumari* and *S. R. Radha*

Chapter 1

**Inaugural address**

# Biodiversity (Bio-Resource) conservation: A global initiative

**Oommen V. Oommen**

Chairman, Kerala State Biodiversity Board, Pallimukku, Pettah, Thiruvananthapuram

All life on Earth is part of one great, interdependent system. It interacts with, and depends on, the non-living components of the planet: atmosphere, oceans, freshwaters, rocks, and soils. Humanity depends totally on this community of life – this biosphere – of which we are an integral part. Biodiversity is the degree of variation of life forms within a given species, ecosystem, biome, or an entire planet. Biodiversity is a measure of the health of ecosystems. Biologists most often define biodiversity as the "totality of genes, species, and ecosystems of a region". An advantage of this definition is that it seems to present a unified view of the traditional three levels at which biological variety has been identified: species diversity, ecosystem diversity, genetic diversity.

Biodiversity is not evenly distributed; rather it varies greatly across the globe as well as within regions. The conservation of biodiversity is fundamental to the success of the development process. As the **Biodiversity of Conservation: a Global Initiative** explains, conserving biodiversity is not just a matter of protecting wildlife in nature reserves. It is also about safeguarding the natural systems of the Earth that are our life-support systems; providing food from the land, Microbes, animals, plants etc and safeguarding the genetic richness on which we depend in the ceaseless struggle to improve our crops and livestock. The Convention on Biological Diversity (CBD) was set up after the Rio Earth Summit in 1992 to ensure the conservation and sustainable use of biodiversity. In 2002, the countries that signed up to the convention promised to reduce the rate of loss of biodiversity by 2010. The COP10 conference was when members of the CBD met again to report back on progress, discuss why the original target has not been met and to make decisions on how to manage biodiversity for the future. COP10, A global Biodiversity conference stands for the tenth meeting of the Conference of Parties to the Convention on Biological Diversity (CBD). It was held in Nagoya, Japan from 18 to 29 October 2010. On 22 December 2010, the UN declared the period from 2011 to 2020 as the UN-Decade on Biodiversity. We should be happy that during October 1-19, 2012, XI Conference of Parties (COP-11), a global mega event on biodiversity, has taken place in Hyderabad, where delegates from 193 party countries met. Pursuant to decision x/46, the eleventh meeting of the Conference of the Parties (COP 11) was held in Hyderabad, India, from 1 to 19 October 2012. COP 11 included a high-level ministerial segment organized by the host country in consultation with the Secretariat and the Bureau. The high level segment took place from 17 to 19 October 2012. The United Nations Decade on Biodiversity serves to support and promote implementation of the objectives of the Strategic plan for Biodiversity and the Aichi Biodiversity Targets. At the Hyderabad conference, The Aichi Targets, a set of 20 specific goals set two years ago in Nagoya, Japan, to halt the rate of biodiversity loss and implement the Strategic plan for biodiversity 2011-2020 of the CBD.

Include the poor in biodiversity conservation: Intelligent management of ecosystems can help to turn local economies around and give destitute households a chance to increase their incomes (Lise Grande, 2012). Key theme of the conference is the impact of biodiversity on the poor. Developed countries agreed to double the fund flow to developing countries by 2015, to halt the rate of loss of the planet's biodiversity, at the Eleventh Conference of the Parties (COP) to the Convention on Biological Diversity (CBD) held at Hyderabad.

Biosafety is a term used to describe efforts to reduce and eliminate the potential risks resulting from biotechnology and its products. The Cartagena Protocol on Biosafety (CPB) is an additional agreement to the Convention on Biological Diversity (CBD) which entered into force in 2003. It aims to ensure the safe transport, handling and use of living modified organisms (LMOs) resulting from modern biotechnology that may have adverse effects on biodiversity, also taking into account risks to human health. CPB established procedures for regulating the import and export of LMOs from one country to another. CPB is reinforced by the Nagoya — Kuala Lumpur Supplementary Protocol on Liability and Redress adopted in Oct 2010 and opened for signatures on 7th March 2011. The Supplementary Protocol specified response measures to be taken in the event of damage to biodiversity resulting from LMOs. Sovereignty principles can rely upon what is better known as Access and Benefit Sharing Agreements (ABAs). The Convention on Biodiversity implies informed consent between the source country and the collector, to establish which resource will be used and for what, and to settle on a fair agreement on benefit sharing.

The concept of 'Biodiversity Hotspots' addresses the dilemma that all conservationists face: what areas are the most immediately important for conserving biodiversity? To qualify as a 'hotspot', a region must contain at least 0.5% species of vascular plants as endemics, and have lost at least 70% of its primary vegetation. The 34 hotspots identified by Conservation International cover 2.3 percent of the Earth's land surface, yet more than 50 percent of the world's plant species and 42 percent of all terrestrial vertebrate species are endemic to these areas. All are threatened by human activities.

Our own organizations are already deeply involved in action to conserve biodiversity. We shall be further developing our own programs in its light. We will be monitoring its implementation and all our own work will reflect the assumption that successful action to conserve the diversity of life on earth is essential for a sustainable human future. The Biodiversity Management Committee (BMC) is a statutory body constituted at LSG level as per the section 20 (1) of the Kerala Biological Diversity Rules, 2008. The section 20 (7) the Rules affirms "the key mandate of BMC as to facilitate the preparation of people's Biodiversity Registers at Grama Panchayat/ Municipality/Municipal Corporation levels. The PBR shall contain comprehensive information on availability and knowledge of local biological resources, their medicinal or any other use or any other traditional knowledge associated with them. The Biodiversity Management Committees and local bodies will be responsible for ensuring the protection of the knowledge recorded in the People's Biodiversity Registers, especially to regulate its access to outside agencies and individuals". Declaration of Biodiversity Heritage Sites (BHS) is the mandate of State biodiversity Boards. Under Section 37 of Biological Diversity Act, 2002 (BDA) the State Government in consultation with local bodies may notify in the official gazette, areas of biodiversity importance as Biodiversity Heritage Sites (BHS). As a preliminary step towards the identification of BHS, KSBB investigated ideal biodiversity rich sites for the declaration and five sites came into the consideration of the Board. Kalasamala, a sacred grove located in Thrissur district harbours critically endangered plant, *Syzigium travancoricum* is one among the five sites identified by the Board.

While our work to conserve bioresources ensures that the vast pool of chemical compounds, genetic resources and other benefits of Earth's species are available to provide new solutions to existing problems, conserving nature also helps us solve problems yet to be discovered. Therefore, we must maintain nature for our own benefit and for the benefit of our descendants.

## Chapter 2

**Special Lecture**

# Bioprospecting bioresources – Issues related to environmental sustainability

**M. N. V. Prasad**
Department of Plant Sciences
University of Hyderabad
Hyderabad, AP, India

**ABSTRACT**

India is one of the richest countries of the world as far as biodiversity is concerned. The Ministry of Environment and Forests (Government of India) during 1991-92 initated a network of botanic gardens to promote ex-situ conservation and propagation of rare plant genetic resources in different ecoregions of the country. The objectives of this programme are i) *Ex-situ* conservation and propagation of rare and endangered endemic species in different ecoregions. ii) Strengthening of the existing infrastructural facilities for conservation of germ-plasms, establishment of seed banks, arboreta and in vitro propagation facilities, iii) Education and public awareness in respect of endemic plant species and ex situ conservation, and iv) Reintroduction of the multiplied redlisted taxa to the natural ecosystems in collaboration with forest department. Under the aegis of this network programme of botanic gardens, we have initiated a " botanic garden project" at the University of Hyderabad to contribute to the identified objectives.

This presentation deals with the medicinal, aromatics and economically important pants of deccan eco-region that are being grown in our Botanical as a measure of *ex situ* conservation. The past traditions and the future forecast of the drug discovery for the dreadful diseases and ailments involves a multifaceted approach combining botanical, phytochemical, pharmacological and molecular techniques. Research on medicinal plants is being hyped for drug discovery and may continue further as the developed nations are keen on phytomedicines. The developing countries are rich in traditional medicines, ethanobotany and medicinal and aromatic plants. Most of the medicinal plants (about 90-95%) are collected from wild and very few are cultivated. Thus, there is an enormous pressure on the plant resources due to escalating demand from the herbal industry. We have initiated our research in the year 2000 on the *ex situ* conservation of red listed medicinal and aromatic plants. A total of 33 redl isted and 55 commonly traded medicinal plants (of different accessions) are being maintained in field gene bank of University of Hyderabad campus. We have succeeded in the rapid multiplication of *Drosera indica, D. burmanii. (in viro); Celastrus paniculatus* by using plant growth promoters. *In vitro* multiplication protocol has been standardized for *Pterocarpus snatalinus*.

Contemporary students of plant sciencences often think plant taxonomy as "*Tax-on-me*". Molecular taxonomy approach would revitalize plant taxonomy owing to the attraction molecular biology as the tools and techniques are same.

Molecular systematic studies have primarily been done in the developed countries. With the rich biological re-sources and available floristic expertise in many developing countries, it will greatly advance the field of systematic and related sub-disciplines if more scientists and students in developing countries are involved in molecular systematic studies along with other integrative systematic investigations.

Biotechnology provides us many new opportunities to make full use of the plant diversity complementing each. The different aspects of biotechnology involve the utilisation of a range of *in vitro* culture systems, protoplasts, cells, callus, embryos, shoots and plantlets. Countries rich in biodiversity (Brazil, Mexico, Chile, China and India) are poor in biotechnology whereas the developed countries life U.S.A., Japan, France, Germany and U.K. are rather poor in biodiversity but rich in biotechnology. The Rio convention on biological diversity (1992) emphasized the biotechnological based conservation of plant diversity in the developing countries.

## Chapter 3

# Conservation of genetic resources and identifying genotypes using molecular markers in *Prunus*: Recent advances in peach and plum

**Jayasankar S., El-Sharkawy I., Sherif S., El-sadr H. and Dinesh Kumar S.**

University of Guelph, Vineland Station, ON, Canada

Perennial crops such as tree fruits have long been identified based on phenotypical traits. However, these identifications may not be entirely accurate and a valuable germplasm that looks similar but genetically different is often lost. Recently there have been tremendous increases in differentiating genotypes that are exhibiting only subtle differences using highly sensitive molecular techniques. Early molecular techniques were largely PCR based such as RAPD, AFLP etc. Later techniques were more sequence based which were able to identify mutations within a gene that can end up in a new genotype. Currently we have techniques that can sequence the whole genome in a relatively short time. These methods have revolutionized the molecular identification that would enable conservation more accurately. These techniques not only will let us look for mutations such as SNPs within the gene as it looks at the whole population, but they might also help us to maintain a proper level of genetic diversity. Such diversity should allow us to sustainably create and access genetic profiles that fit the changing needs. In this presentation we will focus on a relatively new technique called GBS and how we are utilizing GBS to generate molecular identity and use this method for crop improvement in peach among other things. Similarly we will also discuss how a novel mutation in prune plums is tracked down to simple changes in biosynthetic pathways and the practical application of these studies.

RAPD-Random Amplified Polymorphic DNA; AFLP-Amplified Fragment Length Polymorphism; SNP-Single Nucleotide Polymorphism; GBS- Genotyping by Sequencing.

# Chapter 4

# Genetic analysis for conservation and improvement of important wood traits in *Dalbergia sissoo* roxb.

**Ashok Kumar**
Division of Genetics and Tree Propagation
Forest Research Institute
(Indian Council of Forestry Research and Education)
Dehradun, India

## ABSTRACT

*Dalbergia sissoo* Roxb.or Shisham is an important multipurpose tree species of Indian subcontinent and known for its timber qualities. The species is distributed between latitude 21.17° N to 32.60°N, longitude 74.80°E to 93.43°E and altitude up to 900 m in the Sub- Himalayan tracts of India, which occasionally ascends upto 1000 m asl along the river and streams. Though mature trees attain a maximum height of 30 m under favorable conditions with diameter at breast height of 0.8 m, the trunks often found to be crooked leading to higher degree of loses in wood workshops. The genetic improvement of this versatile species has not been taken up in a systematic manner and only sporadic works were carried out merely confined to selection of plus trees and clonal propagation. In fact, there has absolutely been no published record on genetic improvement based on wood parameters, and only limited information available about the degree of genetic variation or the heritability for some wood traits. Nonetheless, it is beyond doubt that the genetics of wood traits is an essential aspect and being a timber species the importance become paramount particularly for degree of variation and its genetic control. Thus the wood analysis for different parameters was carried out in randomly selected twelve clones of a clonal seed orchard of *Dalbergia sissoo* (Roxb.) established during 1997 at Hoshirpur, Punjab, India. The genetic diversity of clones was based on five wood and two growth parameters through Euclidean cluster analysis, which grouped clones into seven clusters. Whereas cluster 1 and 2 contained four and three clones respectively, remaining clusters had just one clone in each. Clone originated from Barielly, Uttar Pradesh (Cluster 7) was found to be the most divergence clone as it maintained highest inter-cluster distance with all other clusters with an exception of cluster 2. Cluster 2, which contained three clones from different geographical regions, also maintained greater inter-cluster distance with other clusters. The divergence analysis has confirmed that the clones planted in the clonal seed orchard are sufficiently divergent and seed harvested from the orchard would maintain high diversity level.

## INTRODUCTION

*Dalbergia* is a pantropical genus with more than 100 species, distributed in different parts of tropical Asia, America and Australia. *Dalbergia sissoo* Roxb. (Shisham) is known internationally for its timber qualities and is one of the fast growing tree species planted in different parts of the world to bridge ever growing gap between demand and supply for timber wood. It is an important multipurpose tree species of Indian subcontinent known for its variety of adaptive and economic traits and commercially production for strong and durable timber. The species is distributed between latitude 21.17° N to 32.60°N, longitude 74.80°E to 93.43°E and altitude up to 900 m in Sub- Himalayan tracts of India, occasionally ascends upto 1000 m asl along the river and streams (Tewari, 1994).

There are no published records are available on genetic improvement based on wood parameters, and only limited information available about the degree of genetic variation or the heritability for wood traits. Nonetheless, it is beyond doubt that the genetics of wood traits is an essential aspect and being a timber species the importance become paramount particularly for degree of variation and its genetic control. It is a known fact that analysis of genetic diversity is a most important component for any of the breeding and genetic improvement programme which basically deals with ecosystem stability and forest sustainability (Libby, 1973). The analysis of genetic diversity becomes still more valuable in tree improvement programme (Zobel, 1971) owing to the long gestation period of trees. Obviously, analysis of genetic diversity enables selection of diverse parents for hybridization either to exploit the gain due to heterosis or synthesize new recombinants for subsequent generations. An attempt has been made to ascertain the magnitude of genetic diversity among the clones of *Dalbergia sissoo* using $D^2$ analysis. The present paper therefore purports towards analysis of genetic diversity of various clones originated from different geographical regions.

## MATERIAL AND METHODS

The wood material was collected from clonal seed orchard (CSO) established in 1996-97 consisting of 27 clones selected from natural distribution range of Shisham of India and Nepal. The clones were planted in randomized block design with a spacing of 5 x 5 m at Punjab (altitude 267.30 m, latitude $31^0$31'36"N and longitude $75^0$48'54"E with rainfall of 1242.20 mm) in an area of 2.40 ha. The CSO was first enumerated fully for height, diameter at breast height (DBH), clear bole height (CBH)), straightness and branching behavior. Thereafter, the index values for individual ramets and clones were established, and a total of twelve clones obtained maximum index value were selected for wood parameters. Further, two maximum scoring ramets for selected clones were marked in all the replications individually of CSO for analysis of wood parameters. Five mm thick increment borer was used for collecting wood samples at breast height (1.37 m) and in this manner a total of 72 samples were collected. The cores were rubbed with formalin to prevent the fungal attack. Each core was divided into 4 equal segments from pith to periphery and named as pith, inner, middle and outer segments.

In this way, a total of 288 samples were prepared for wood analysis and standard laboratory procedures were followed for micro-slides and macerations. The green volume of the samples was determined using water displacement method and the dry matter of the samples was calculated by oven drying at 103±$2^0$C till constant weight. The specific gravity of different samples was accordingly calculated.

The radial cores were macerated with 50% nitric acid and a pinch of potassium chlorate as per Schultz's method. In fact, each core was individually dipped into test tubes filled with above mixture and kept in the sun light for maceration. The solution was decanted and washed several times with water to remove traces of acid and microscopic observations obtained. The average tangential diameter of the vessels was determined from 25 measurements and same numbers of observations were recorded for various other parameters of the macerated wood for four locations *viz.* pith, inner, middle, and outer.

The growth and wood data were analyzed statistically as described by (Sukhatame and Amble, 1989) and $D^2$ analysis was performed using Mahalanobis's $D^2$ statistics (Mahalanobis, 1936) using SAS (version 9.1.2 software for windows).

## RESULTS AND DISCUSSION

The analyzed data of growth and wood characters depicted that on the basis of Euclidean distances dendrogram, twelve clones of *Dalbergia sissoo* were grouped into 7 clusters (Fig. 1).

**Figure 1: Dendrogram based on Tocher method**

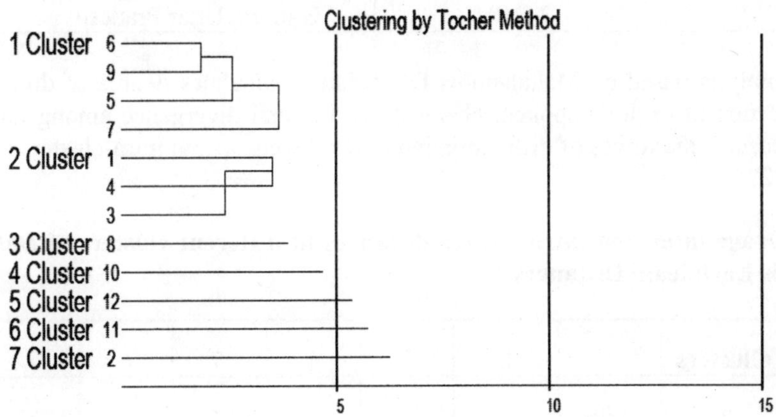


Cluster 1 consisted 4 clones (clones 41, 192, 33 and 57) followed by cluster 2 with 3 clones (Clone 2, 19 and 12) and rest of the clusters consisted one clone each (Table 1). The clustering indicated that though the species had sufficient genetic diversity, clones selected from the same geographical region came together in the same cluster (Table 1).

All the clones of Cluster 1 and Cluster 2 are selected from the same geographical region of Pathri (except clone 192) and Bijnore, respectively. The position of clone 198 and clone 10 in cluster 3 and 7 again indicates that there is enough diversity within the region as these clones tend to move to different clusters rather than group with cluster 1 and 2. Similarly, clones selected from the geographical region of Gonda (232 and 204) also grouped into two different clusters (Cluster 5 and 6). It is thus obvious that clones selected from a single region were spread over different clusters which could probably be due to heterogeneity of genotypes, genetic architect of different genotypes, higher intensity of selection and high degree of general combining ability (Singh, 2007).

**Table 1- Constituents of different clones and clusters**

| Cluster | S. No. (clone) | Origin |
|---|---|---|
| 1 | 6 (41) | Pathri,Uttarakhand |
| | 9 (192) | Saharanpur, Uttar Pradesh |
| | 5 (33) | Pathri, Uttarakhand |
| | 7 (57) | Pathri, Uttarakhand |
| 2 | 1 (2) | Bijnore, Uttar Pradesh |
| | 4 (19) | Bijnore, Uttar Pradesh |
| | 3 (12) | Bijnore, Uttar Pradesh |
| 3 | 8 (66) | Pathri, Uttarakhand |
| 4 | 10 (198) | Bareilly , Uttar Pradesh |
| 5 | 12 (232) | Gonda , Uttar Pradesh |
| 6 | 11 (204) | Gonda , Uttar Pradesh |
| 7 | 2 (10) | Bijnore, Uttar Pradesh |

Divergence analysis based on Mahalanobis $D^2$ statistics calculates degree of diversification and relative proportion of each component character to the total divergence among total individuals studied. It measures the forces of differentiation at two levels, *i.e.*, at intra-cluster and inter cluster levels (Table 2).

**Table 2: Average inter and intra cluster distances in different clusters of *Dalbergia sissoo* (Mahalanobis Euclidean² Distances)***

| Clusters | 1 | 2 | 3 | 4 | 5 | 6 | 7 |
|---|---|---|---|---|---|---|---|
| 1 | **4.06** | | | | | | |
| 2 | 28.21 | **3.61** | | | | | |
| 3 | 6.26 | 29.69 | **0.00** | | | | |
| 4 | 9.82 | 24.00 | 8.47 | **0.00** | | | |
| 5 | 15.73 | 38.54 | 9.91 | 5.35 | **0.00** | | |
| 6 | 10.55 | 45.02 | 5.72 | 10.55 | 6.96 | **0.00** | |
| 7 | 43.89 | 10.24 | 50.06 | 43.24 | 68.07 | 74.44 | **0.00** |

*Figure in bold indicate intra-cluster distance

The intra-cluster distance ranged from 0.00 (cluster 3, 4, 5, 6 and 7) to 4.06 (cluster 1) to indicate substantial diversity among the clones studied. The inter cluster $D^2$ values ranged from 5.35 to 74.44, maximum being between cluster 7 and 6 (74.44) followed by cluster 7 and 5 (68.07), 7 and 3 (50.06), 6 and 2 (45.02) and 7 and 1 (43.89) as described in Table 2 and Fig 2.

**Figure 2: Mahalanobis Euclidean² Distances**

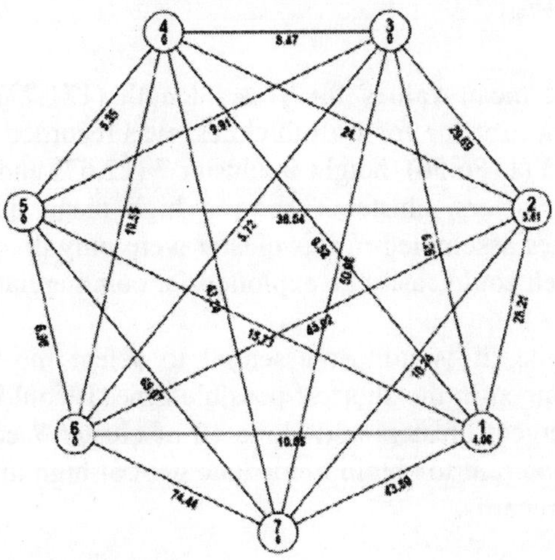

In fact, cluster 7 (clone 10) showed maximum inter cluster diversity from all other clusters (except cluster 2) and the level of diversity was also substantially higher. Similar pattern of genetic diversity was observed in *Casuarina equisetifolia* by grouping 42 clones into 12 clusters (Kumar and Gurumurthi, 2000). Singh (1993) explained that maximum heterosis occurs at an optimal or intermediate level of divergence. Besides the selection of divergent parents for hybridization and establishment of seed orchards, this technique measures the degree of diversification and determines the relative proportion of each component character to the total divergence. In forest tree crops, genetic divergence studies are also an effective tool for establishing seed orchards with diverse parents so that improved seed in most economic manner is harvested as diverse parents would have equal opportunities for hybridization and production of quality seeds (Kumar and Gurumurthi, 2000). Hybridization between the clones selected from diverse clusters is expected to express higher heterosis and produce desirable recombinants among the transgressive segregants (Dhillon *et al.*, 2009). The technique has effectively been applied in various forest tree species to find out distinct populations / provenances / progenies in *Eucalyptus camaldulensis*, (Burley et al., 1971; Andrew, 1973), *nilotica* (Bagchi, 1992), *Gmelia arborea* (Kumar, 2007), *Dalbergia sissoo* (Kumar *et al.*, 2012) and *Leucaena leucocephala* (Kumar *et al.*, 2013).

**Table 3: Cluster means for various characters of twelve clones of *Dalbergia sissoo***

| Cluster No. | Characters | | | | | | |
|---|---|---|---|---|---|---|---|
| | WL | VL | VD | FL | DBH | H | Sp.Gr |
| 1. | 3.92 | 171.77 | 223.11 | 1003.83 | 1038.38 | 22.08 | 63.68 |
| 2. | 3.29 | 164.45 | 192.03 | 954.76 | 999.00 | 17.83 | 65.35 |
| 3. | 3.96 | 168.01 | 200.32 | 985.57 | 1100.00 | 22.50 | 69.09 |
| 4. | 3.81 | 162.27 | 189.97 | 959.22 | 922.00 | 20.83 | 57.49 |
| 5. | 3.86 | 160.65 | 194.15 | 993.37 | 1186.50 | 21.33 | 57.26 |
| 6. | 4.07 | 168.73 | 201.78 | 988.09 | 1115.00 | 18.67 | 62.74 |
| 7. | 3.24 | 168.39 | 193.98 | 910.85 | 1038.00 | 22.67 | 62.76 |

The cluster means for different characters under study revealed considerable genetic differences among the clones. (Table 3).

Cluster 1 had the highest mean values for vessel length (171.77), vessel diameter (223.11). However, the maximum mean value for wall thickness was recorded in cluster 6 (4.07), diameter at breast height in cluster 5 (1186.50), height in cluster 7 (22.67) and specific gravity in cluster 3 (0.69). Among different clusters, cluster 1 showed high mean performance for most of the characters. Hence the clones assembled in this cluster were only divergent had promising growth and wood parameters, which could easily be exploited for commercially.

In hybridization programme, it is utmost essential to select most divergent parents so that maximum heterosis is obtained in the shortest possible time. It could easily be done by selecting the parents from two divergent clusters and clone 10 of cluster 7 could be used in a variety of combinations. It is also important to obtain immediate gain of high diversify by establishing these clones in the clonal seed orchards.

## CONCLUSION

Clonal forestry has mostly been alleged to offer narrow genetic base, yet vegetative propagation has strongly been recommended (Camphinhos and Ikemori, 1980; Ahuja and Libby, 1993). Adding a large number of intensively selected clones in the clone bank / trials and determining the diversity as well as genetic distance would immensely help to maintain genetically diverse population of high yielding clones. Using the planning stock from such gene bank, it is possible to develop a breeding programme with selective individuals of desired constitution. It is emphasized that clonal strategy should form a part of overall genetic improvement programme. However, the inadequate progress made in genetic improvement need not per se limit clonal strategy if the selection is carried out from diverse population by adopting proper tools and techniques so that genetically broad based productive clones are maintained (Gurumurthi *et al.*, 1994)

## REFERENCES

Ahujha MR, Libby W. (1993). Clonal Forestry : Genetics and Biotechnology. Springer-Verlag, New York. P 227

Andrew IA (1973). *Variation in leaf morphology among provenances of Eucalyptus camaldulensis Dehn. grown in Rhodonsia.* Agric. Res. **11**: 159-169

Bagchi SK (1992). *A preliminary study on the genetic divergence of Acacia nilotica through seed parameters.* Indian Forester **118** (6): 416-424

Burley J, Wood PJ, Hans AS (1971). *Variation in leaf characteristics among provenances of Eucalyptus camaldulensis Dehn. grown in Zambia.* Aust. J. Bot. **19**: 237-249

Camphinhos E, Ikemori YK (1980). *Mass production of Eucalyptus species by rooting cuttings.* **In**: Proc: IUFRO Symposium on genetic improvement and propagation of fast growing trees. Sao Pede, Sao Paulo, Brazil, 16p

Dhillon RS, Verma RC, Dhanda SK, Sheokand R, Kumari S (2009). *Genetic divergence based on quantitative variation for some seed traits in plus trees of neem (Azadirchta indica A. Juss*). **Indian.** J. of Agro-Forestry. **11**(1): 55-60.

Gurumurthi K, Yasodha R, Ravi N, Rani M, Kumar A, Verma RK (1994). Biotechnological option in the tree farming, edited by Thumpan PK, Peekay Tree Crop Development Foundation, Cochin , India, pp 393-423.

Kumar, A (2000). *Growth performance and variability in different clones of Gmelina arborea.* Silvae Genetica, 56 (1): 32-36.

Kumar A, Gurumurthi K (2000). *Genetic divergence studies on clonal performance of Casuriana Equstifolia.* Silvae Genetica, 49(2): 57-60

Kumar A, Bhatt A, Ravichandran S, Pande P K and Dobhal, S (2012). *Genetic divergence for growth and wood parameters in different clones of Dalbergia sissoo Roxb.* Journal of Forestry Research, 23(2): 285-288.

Kumar A, Pande P K, Ravichandran S, Naithani S, Kishore P B K, Raturi A, Gautam P, Dobhal S and Sharma P (2013). *Genetic analysis of growth and wood variations in Leucaena leucocephala.* **Journal** of Forestry *Research,* 24(1).485-493

Libby, WJ (1973). *Domestication strategies for forest trees.* Canadian Journal of Forest Research, 3: 265-276

Mahalanobis, P C (1936). *On the generalized distance in statistics.* Proc., Nat. Ins. Sci., India. 2: 49-55

Parrotta, JA (1989). *Dalbergia sissoo* Roxb. SO-ITF-SM-24. USDA Forest Service, Rio Piedeas, Puerto Rico, 5p

Singh, BD (2007). Plant Breeding:Principles and methods, Kalyani Publishers, New Delhi. Pp. 667

Sukhatme, PV, Amble VN (1989). Statistical methods for agricultural workers. Publication Division, Indian Council of Agricultural Research, New Delhi 110 012. Pp. 359

Tewari, DN (1994) A monograph on *Dalbergia sissoo* Roxb. International Book Distributor, Dehradun, 316p

Thothath,ri K (1987). Taxonomic revision of the tribe *Dalbergieae* of Indian sub-continent. BSI Howrah

Troup, RS (1921). The Silviculture of Indian Trees 1(Rev. Edition 1980). Oxford Press, Oxford

Zobel, BJ (1971). The Genetic Improvement of Southern Pines. Sci. Amer. **225**: 94-103

# Chapter 5

## Special Lecture

# Mechanism of salt exclusion and secretion in the mangrove *Avicennia*

**Pannaga Krishnamurthy, Pavithra AJ, Wee-KeeTan, Teck-Kwang Lim, Tit-MengLim, QingsongLin, Chiang-Shiong Loh and Prakash P. Kumar***

Department of Biological Sciences, National University of Singapore, Singapore

**ABSTRACT**

Mangroves thrive in the high salt environment of the intertidal coastal regions. They are continually subjected to high salt stress due to the presence of $Na^+$ and $Cl^-$as the predominant ions in seawater. They have evolved adaptive survival strategies such as salt exclusion at the root and salt secretion from the leaves through specialized salt glands. During water intake the salt excluding mangrove species are known to filter up to 95% of the sodium from external medium while, the salt secretors can filter only about 85% of the external sodium (Scholander *et al.*, 1962).Because sea water has approximately 500 mM salt, 15% of the sodium which enters the xylem sap is still toxic to the plants and therefore the excess salt needs to be removed from the leaves of the salt secretors. The mechanism of salt exclusion/filtration at the root level and secretion though salt glands in the leaves are still not well characterized. Using the salt secreting mangrove *Avicennia officinalis*we initiated physiological and proteomic studies aimed at understanding the mechanism of salt secretion.

We examined the process of secretion of salt solution in the leaves and were able to demonstrate that the process is rhythmic with increases and decreases in volume of salt solution secreted from each salt gland. This suggests that the process is not a one way secretion of solution, but involves conservation of the secreted water by reabsorbing some of it. The hydrophobic endodermis and exodermis in the root tissues help to change the bulk flow into symplastic route for the uptake of water and dissolved salts in to the xylem, thereby playing a major role in root ultrafiltration. We studied the effect of salt stress on the formation of root hydrophobic barriers using suberin-specific stains. We observed differential development of suberin layer in the root (endodermis and exodermis). Subsequently, we could demonstrate the root filtration efficiency to be about 85% in

*Avicennia* using the pressure bomb techniqueto estimate salt filtration efficiency and bulk flow in *Avicennia* seedlings.

In order to better understand the mechanism underlying he salt secretion process, we attempted to identify the key transporters in the membranes of the salt gland cells and surrounding cells. By proteomics analysis of *Avicennia* leaves we identified several transport proteins located in the plasma membrane and the tonoplast (vacuolar membrane).The key transporters identified by mass spectrometry will be chosen for further molecular studies, because they may be involved in transporting ions. The results obtained so far from our studies will be discussed.

## ACKNOWLEDGEMENTS

This work was supported by a research grant from Singapore National Research Foundation under its Environmental & Water Technologies Strategic Research Program administered by the EWI of the Public Utilities Board (PUB), Singapore, NRF-EWI-IRIS (2P 10004/81) (R-706-000-010-272).

## REFERENCES

Fahn A, Shimony C. (1977). *Development of the glandular and non-glandular leaf hairs of Avicennia marina* (Forsskil) Vierh. Botanical Journal of the Linnean Society 74, 37-46.

Scholander P. F, Hammel HT, Hemmingsen E, Garey W. (1962). *Salt balance in mangroves.* Plant Physiology 37, **722-729.**

Tomlinson P.B. (1986). The botany of Mangroves. London, UK: The Cambridge University Press.

Yeo A.R. (1983). *Salinity resistance: physiologies and prices.* Physiologia Plantarum 58, 214-222.

Chapter 6

Special Lecture

# Conservation strategies for plant genetic resources with special reference to agro-biodiversity in India

**N. Sivaraj**
National Bureau of Plant Genetic Resources (NBPGR), Regional Station
Hyderabad, AP, India

## INTRODUCTION

Plant genetic resources are one of the most important components of biodiversity and these hold the key to foundation of agriculture, food and nutritional security. The importance of plant genetic resources has increased significantly in the recent years with the changing global scenario in material ownership and the legal regimes with respect to access to genetic resources under the International Agreements. The sustainable management of plant genetic resources is the major concern in today's world, as ever increasing population and rapid technological growth are putting tremendous pressure on these resources. Genetic resources management involves a range of activities including plant exploration and collecting, conservation, characterization and evaluation, exchange of germplasm, genetic enhancement. Modern molecular approaches can, and increasingly so, contribute to all these activities. It is now well understood that advances through biotechnologies indeed are possible only through an increased access to wide range of plant genetic resources. Indian agriculture production scenario is also being confronted by numerous other problems, which calls upon national strategic approaches to be adopted with respect to germplasm exchange for broadening the genetic base, conservation of PGR, exploring and promoting the under-utilized crops; developing genetic diversity to reduce crop vulnerability to climate changes and to overcome other yield limiting characters. Conservation strategies for the plant genetic resources with a special reference to agro-biodiversity are being highlighted.

## STATUS OF INDIA'S AGROBIODIVERSITY

The Indian gene centre is among the 12 megadiversity regions of the world. About 25 crop species were domesticated in India. It is endowed with rich diversity of more than 18,000 species of higher plants including, 160 major and minor crop species and 325 of their wild relatives. Around 1,500 wild edible plant species are widely exploited by native tribes. These include 145 species of roots and tubers, 521 of leafy vegetables/greens, 101 of buds and flowers, 647 of fruits and 118 of seeds and nuts. In addition, nearly 9,500 plant species of ethno botanical uses have been reported from the country, of which around 7,500 are for ethno medicinal purposes and 3,900 are multipurpose/edible species (NAAS, 1998).

## Conservation of plant genetic resources

The proper conservation and use of plant genetic resources which is also a component of agro-biodiversity is fundamental to sustainable development and food security. As plant genetic resources are fundamental to the fulfillment of basic needs of human community like food, fodder, fuel, fibre etc., their loss has serious implications on economic and social development. Hence, plant genetic resources have to be conserved and managed for sustainable use.

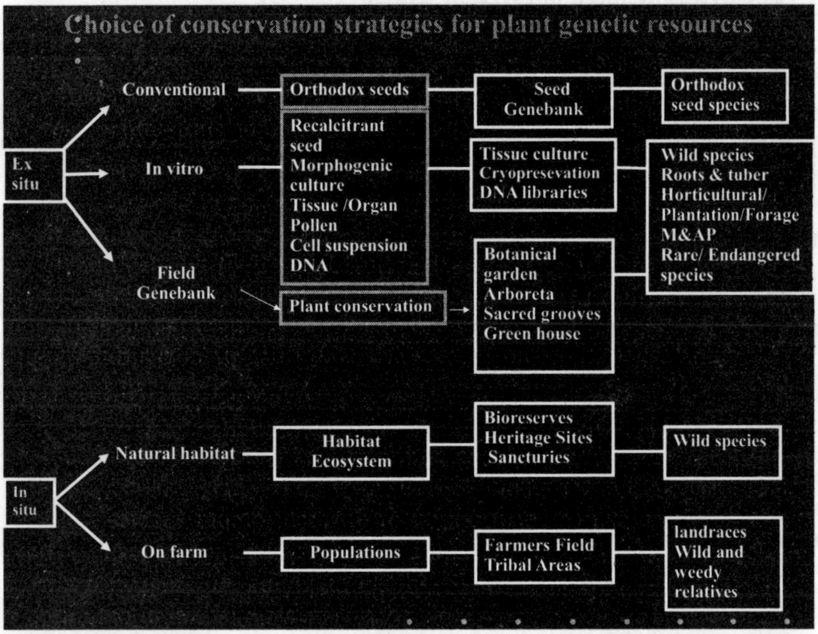


(*Source: NBPGR, New Delhi*)

Germplasm is the basic material for the plant breeding or crop improvement. Realizing the importance of the germplasm, systematic collection and conservation of germplasm is being carried out for all the crop species worldwide (Pillai *et al.*, 2002). Biotechnological tools are currently being well utilized in the germplasm management to identify specific accessions, assess the genetic diversity among the core collections, confirm the true hybridism, and evaluate the taxonomical status and to detect duplicates in the core germplasm collections.

## *Ex-situ* conservation

It refers to conservation of genetic material of biological diversity outside their natural habitat in facilities supporting either storage or perpetuation with conditions suited to maintain their viability and genetic constitution. *Ex-situ* conservation virtually safeguards and provides required supply of germplasm for research and breeding (Singh et al., 2004). Some of the possible approaches can be grouped in:

**Plant conservation**: (a) Botanical garden, (b) Arboreta, (c) Herbal garden, (d) field genebank and (e) clonal repositories.

**Seed conservation**: (a) Low temperature storage of orthodox seeds (seed genebank), (b) Cryopreservation: storage of orthodox, intermediate and recalcitrant (embryonic axis) seeds in liquid nitrogen at -150 to -196 °C. The status of conservation of plant genetic resources in the National Gene Bank located in New Delhi is provided in Table 1.

***In-vitro* conservation:** (a) Conservation of cells, tissues, organs in glass or plastic containers under aseptic conditions through slow growth of cultures (b) Cryopreservation of cultures (tissues, organs, pollen or cultures in liquid nitrogen at -150 to -196 °C).

**DNA conservation:** The basic objective in conservation of PGR is conservation of genetic diversity existing in the form of a functional unit called 'gene'. The whole genome in the form of genomic library or a sequence of DNA in the form of DNA library may be conserved following the appropriate conservation method. A recent initiative in this regard is the National Genomic Resources Repository established at NBPGR, New Delhi.

## *In-situ* conservation

It refers to conservation of genetic resources within their ecosystem and natural habitats. In-situ conservation is required for forestry species, species belonging to complex ecosystem, endangered, and wild relatives of crop plants etc. It has mainly two approaches (a) ecosystem approach (biosphere reserves) and (b) habitat approach (sacred groves, gene sanctuaries, national parks). *In situ* conservation involves continued habitat maintenance. In some contexts *in situ* methods can directly address the causes of environmental degradation and seek to limit their effect.

**Table 1. *Ex-situ* conservation status of Plant Genetic Resources in the National Gene Bank, India**

| STATUS OF BASE COLLECTIONS IN NATIONAL GENEBANK (-18 °C) (As on 31st October, 2012) | |
|---|---|
| Paddy | *93508* |
| *Wheat* | *40044* |
| *Maize* | *9479* |
| *Others* | *12010* |
| **Cereals** | **155041** |
| *Sorghum* | *20414* |
| *Pearl millet* | *8395* |
| *Minor millet* | *22303* |
| *Others* | *5283* |
| **Millets and forages** | **56395** |
| *Amaranth* | *5558* |
| *Buckwheat* | *880* |
| *Others* | *387* |

| | |
|---|---|
| **Pseudo Cereals** | **6825** |
| *Chickpea*<br>*Pigeonpea*<br>*Mung bean*<br>*Others* | *16894*<br>*11427*<br>*3704*<br>*26135* |
| **Grain Legumes** | **58160** |
| *Groundnut*<br>*Brassica*<br>*Safflower*<br>*Others* | *14603*<br>*10602*<br>*8048*<br>*24118* |
| **Oilseeds** | **57371** |
| *Cotton*<br>*Jute*<br>*Others* | *6815*<br>*2914*<br>*2211* |
| **Fibre Crops** | **11940** |
| *Brinjal*<br>*Chilli*<br>*Others* | *4084*<br>*2011*<br>*18967* |
| **Vegetables** | **25062** |
| *Custard apple*<br>*Papaya*<br>*Others* | *59*<br>*23*<br>*448* |
| **Fruits** | **530** |
| *Opium poppy*<br>*Ocimum*<br>*Tobacco*<br>*Others* | *350*<br>*465*<br>*1483*<br>*4472* |
| **Medicinal & Aromatic Plants & Narcotics** | **6770** |
| *Coriander*<br>*Sowa*<br>*Others* | *701*<br>*91*<br>*2555* |
| **Spices & Condiments** | **3347** |
| *Pongam oil tree*<br>*others* | *395*<br>*2048* |
| **Agro-forestry** | **2443** |
| *Lentil*<br>*Pigeonpea* | *7712*<br>*2523* |
| **Duplicate Safety Samples** | **10235** |
| **Total** | **394119*** |

***The figure includes 4103 Released varieties and 2198 Genetic stocks; No. of crop species conserved – 1584** *(Source: www.nbpgr.ernet.in)*

## *In-vitro* technologies for conservation of PGR

During the past few decades, a great deal of interest has been shown to utilize biotechnological approaches for the management of plant genetic resources, particularly conservation. The reason for the upsurge of this interest is to look for a sustainable alternative to conventional approach of germplasm conservation in case of recalcitrant seed species as well as vegetatively propagated crop plants. Use of *in vitro* technologies can overcome the problems of medium-term conservation of PGR, and cryopreservation offers the solution for their long-term conservation (Mandal and Rekha Choudhury, 2004). In fact, for efficient management of PGR, a wide range of biotechnological methods are now available including tissue culture, cryopreservation, biochemical and molecular genetic marker analyses, immunological diagnosis etc.

*In-vitro* conservation strategies can be technically divided into two categories (1) *In vitro* active genebank (IVAG) or tissue culture repository and (2) *In vitro* base genebank (IVBG) or cryobank. The former genebank is for the short to medium term storage of germplasm using shoot cultures or plantlets under slow or normal growth procedures but the cultures are maintained under growing state. While cryobank is meant for the long term conservation PGR, usually using embryos (somatic or zygotic), shoot tips and pollen under suspended growth.

**Storage using *in vitro* slow growth:** The various technologies that have been used for maintenance of *in vitro* collections under slow growth include the use of reduced temperature and reduced light conditions, use of minimal growth media, use of growth retardant etc. (Engelmann, 1991).

**Storage using *in vitro* normal growth:** Some of the species (*Musa*, Yams, *Piper, and Alocasia*) can be maintained on their respective simple shoot culture media for 10-12 months at 25°C without being subjected to growth inhibitory treatments. Alteration of culture conditions and/or culture media helped in the induction of storage organs in several species, which in turn increased the shelf life of cultures at 25°C. However, management of large collections in vitro is problematic even if the subculture intervals are greatly extended. As the germplasm is conserved under growing conditions, the stress imposed may have deleterious effects like induction of variation, loss of regenerative capacity etc. in the long run.

## Storage using cryopreservation

Cryopreservation, that is the non-lethal storage of biological material at ultra-low temperature, usually that of liquid nitrogen (-196 °C) is the only method currently available for the long-term conservation of germplasm of the problem species. At this temperature, all cellular divisions and metabolic processes are arrested. The plant material can, thus, be stored without alteration or modification of a theoretically unlimited period time. Moreover, cultures are stored in a small volume, protected from contamination with very limited maintenance. Cryopreservation has been applied to seeds of more than 300 orthodox species (Chaudhury, 2000) and pollen of several species (Towill and Walters, 2000). The status of germplasm conserved under *in-vitro* bank and cryo bank at NBPGR, New Delhi is provided in Table 2.

Cryopreservation has been applied for the conservation of secondary product forming cells (industrial cultures) such as *Daucas carota* (Anthocyanin), *Panax ginseng* (Ginsenoside), *Dioscorea composita* (Diosgenine), *Coleus blumeri* (Rosmarinic acid), *Lavandula vera* (Biotin),

*Digitalis purpurea* (Glycosylation, Beta-hydroxylation), *Catharanthus roseus* (Indole alkaloids), *Berberis wilsoniae* (Berberine), Beta vulgaris (Betaxanthine, betacyanin) etc. which are having very high industrial value.

## Policies, legislation acts related to PGR

The major global issues impacting genetic resources management include the CBD, ITPGRFA, WIPO, CITES, TRIPS and SPS. Recognizing the value of plant genetic resources and in harmony with several international treaties, Indian government also has come out with Biological diversity act (BDA) and Plant varieties protection and farmers' rights act (PVP&FRA) providing unique rights to farmers on par with qualified breeders. The Patent Act was also modified suitably ensuring prior approval of the National Biodiversity Authority before award of the patents involving biological resources.

**Table 2. Tissue culture and cryo preservation of agro-biodiversity at NBPGR, New Delhi (*As on November 2012*).**

| Crop Group | Present Status |
|---|---|
| ***In vitro* bank** | |
| 1. Tropical fruits | 416 |
| 2. Temperate and minor tropical fruits | 327 |
| 3. Tuber crops | 618 |
| 4. Bulbous crops | 171 |
| 5. Medicinal & aromatic plants | 174 |
| 6. Spices and industrial crops | 380 |
| **TOTAL** | **2086** |
| ***Cryobank*** | |
| 1. Recalcitrant | 0 |
| 2. Intermediate | 5772 |
| 3. Orthodox | 3346 |
| 4. Dormant bud (Mulberry) | 380 |
| 5. Pollen (Mango) | 343 |
| **TOTAL** | **9841** |

*(Source: www.nbpgr.ernet.in)*

Provisions to protect and conserve threatened plant species through notification is implemented through BDA. A Plant quarantine order was issued under Destructive insect pests act facilitating safe import of plant genetic resources through NBPGR for research purposes. New guidelines are also issued by the Department of Biotechnology for safe exchange of genetically modified plants. Indian Council of Agricultural Research has instituted a mechanism for registration of experimentally developed superior germplasm of potential value.

## CONCLUSION

Future prospects are highly encouraging in terms of the development and application of new techniques and protocols within the context of germplasm conservation. *In situ* methods have been developed for ecosystem and habitat conservation, and forest ecosystems and species are preserved *in situ* through forest reserves and managed use. Since the beginning of the modern conservation effort for PGR, *in situ* conservation strategies have been identified as important elements in the overall conservation effort. Thus, conservation of plant genetic resources is vital as it leads to increased productivity, food security, robust and stable farming systems and ecosystems in the country.

## REFERENCES


Choudhury, R. (2000). *Cryopreservation of seeds, embryos, embryonic axes and pollen at national cryobank of NBPGR.* In: Cryopreservation of Tropical Plant Germplasm (Eds. Engelmann, F and Takagi, H.). JIRCAS, Tsukuba, Japan/IPGRI, Rome Italy 457-459.

Engelmann, F. (1991). *In vitro conservation of tropical plant germplasm: a review.* Euphytica 57: 227-243.

Hodgkin, T. and Ramanatha Rao, V. (2001). People, Plants and DNA: *Perspectives on the scientific and technical aspects of conserving and using plant genetic resources. Managing plant genetic diversity.* In: Proceedings of the international conference on science and technology for managing plant genetic diversity in the 21st century (Eds.Engels, J., Ramanatha Rao, V., Brown, A.H.D. and Jackson, M.). CAB International, Wallingford.

Joshi Swati P., Prabhakar K., Ranjekar and Vidya SG. (1999). *Molecular markers in plant genome analysis,* Curr.Sci., 77: 230-240.

Mandal, B.B. and Rekha Choudhury (2004). *Ex-situ approaches for conservation plant genetic resources: achievements and challenges.* In: Plant genetic resource management (Eds. Dhillon, B.S., Tyagi, R.K., Arjun Lal and Saxena, S.). Narosa Publishing House, New Delhi, pp 194-203.

NAAS (1998). *Conservation, management and use of agrobiodiversity. Policy paper 4.* National Academy of Agricultural Sciences, New Delhi. 7 p.

Pillai, S.V., Sunderasan, S. and Sheela, M.N. (2002). *Biotechnology in cassava germplasm conservation and breeding in India.* In: Cassava Research and Development in Asia: Exploring New Opportunities for an Ancient Crop, Proceedings of the 7th regional cassava workshop, Oct-28 to Nov-1, Bangkok, Thailand, 140-149.

Ramanatha Rao, V. (2004). *Management of plant genetic resources at international level for a food secure future.* In: Plant genetic resource management (Eds. Dhillon, B.S., Tyagi, R.K., Arjun Lal and Saxena, S.). Narosa Publishing House, New Delhi, **pp. 361-374.**

Singh, A.K., Tyagi, R.K. and Kameshwar Rao, N. (2004). *Ex-situ approaches for conservation plant genetic resources: achievements and challenges.* In: Plant genetic resource management (Eds. Dhillon, B.S., Tyagi, R.K., Arjun Lal and Saxena, S.). Narosa Publishing House, New Delhi. pp 179-193.

Towill, L.E. and Walters, C. (2000). *Cryopreservation of pollen.* In: Cryopreservation of Tropical Plant Germplasm (Eds. Engelmann, F and Takagi, H.). JIRCAS, Tsukuba, Japan/IPGRI, Rome Italy, pp 115-129.

Chapter 7

# Orchids in the adjoining areas of Shimla, Himachal Pradesh: Distribution, survival threats and some conservation measures

**Ravi Kant and *Ashli R.**
Department of Botany, Shoolini Institute of Life Sciences and Business Management
Solan – 173212 (H.P.), India
*Department Of Botany, P.S.G.R.Krishnammal College for Women, Coimbatore, Tamilnadu

**ABSTRACT**

Orchids are known for their beautiful flowers of myriad shape, size and colour. Orchids contain a variety of alkaloids, flavonoids, phytochemicals and used in the traditional system of medicines to cure a variety of ailments. During March 2009-Octber 2012, we have explored the distribution and present status of different orchids near Shimla including Summer Hills, way to Tara Devi Temple, Fagu and Narkanda. During the survey, 11 orchid species representing different subfamilies were reported. However, due to many anthropogenic factors, their natural population is decreasing every year. If the factors continue, these orchids may become extinct in the near future. Presently, some conservation measures have been suggested.

## INTRODUCTION

Medicinal importance of plants has been realised even from the time of early man. Orchids belong to one of the largest and highly evolved families of angiosperms called Orchidaceae with nearly 25, 000 to 35,000 species under 750 to 850 genera distributed all over the world. In India, there are about 1600 orchid species (Medhi and Chakrabarti, 2009). Orchids are the rich source of alkaloids, glycosides and phytochemicals and used in the traditional system of medicine like Ayurveda, Siddha, Yunani, Homeopathy, Traditional Chinese Medicine (TCM) etc. to cure various ailments. Recently some potential drugs have been isolated from orchids Hossain *et al.*, 2013). Orchids are pollinators specific and require particular mycorrhizal association in their life cycle. Decrease in the number of pollinators, fungal diversity and anthropogenic factors like construction work, habitat destruction, urbanisation, illegal trade and introduction of exotic species have significantly reduced the orchid diversity from their natural habitats. Himachal Pradesh is situated in the north-western Himalayas of India between latitudes 30°22′44″N -

33°12'40"N and longitude 75°45'55"E - 79°04'20"E. The state is rich in orchid diversity with 83 species (Vij *et al.,* 2011). Adjoining areas of Shimla including way to Tara Devi Temple, Kufri, Fagu and Narkanda are well known tourist's places of Himachal Pradesh. A field survey was conducted in these areas to explore distribution of different orchid species in these places, their present status and survival threats to them.

## MATERIAL AND METHODS

Identity of the presently reported orchid species was confirmed by consulting the literature (Deva and Naithani, 1986). Periodic field surveys were conducted in adjoining areas of Shimla i.e., way to Tara Devi Temple, Kufri, Fagu and Narkanda between March 2009-Oct. 2011 to locate different orchids and notes were prepared on their distribution and survival threats.

## RESULTS AND DISCUSSION

In the selected areas, 11orchid species belonging to different subfamilies were reported. The names of orchid species, their distribution in the selected areas, flowering period and their uses have been presented in the table: 1.

**Table: 1**, Names of orchid species, their distribution in the selected areas, flowering period and their uses

| S. No. | Botanical name | Distribution | Flowering period | Uses |
|---|---|---|---|---|
| 1 | *Cypripedium cordigerum* D. Don | Kufri | March-April | Powder made from rhizome is used to get relief from spasm and as nerve stimulant (Khan et al., 2013) |
| 2 | *Goodyera biflora* (Lindl.) Hook. f. | Taradevi | May-June | Uses as pot plant. |
| 3 | *Goodyera repens* (L.) R. Br. | Fagu | May- June | It is used in the preparation of Chyavanprash, as stomach ache, to treat cold, kidney disorders, female disorders, toothache and urination problems . |
| 4 | *Spiranthes sinensis* (Pers.) Ames. | Fagu | July-August | It is an important medicinal herb in the Ayurvedic system of medicine. The roots are used as aphrodisiac and to cure various diseases like haemoptysis, headache, meningitis and chronic dysentery. |
| 5 | *Satyrium nepalense* D. Don. | Tara Devi | July-August | Tubers are sweet in taste and used as food. It is also used to cure malaria and dysentery (Chauhan, 1990). Local people of Uttrakhand used the tubers of plant to prepare energetic tonic and to cure fever |

| S. No. | Botanical name | Distribution | Flowering period | Uses |
|---|---|---|---|---|
| | | | | (Mishra and Saklani, 2012). |
| 6 | *Herminium lanceum* (Thunb. ex Sw.) Vuijk | Tara Devi | July-August | It is use to cure urine related problems. |
| 7 | *Neottia listeroids* Lindl. | Tara Devi | July-August | ------------ |
| 8 | *Epipactis helleborine* (L.) Crantz | Fagu | March-April | Roots are used in insanity . |
| 9 | *Malaxis muscifera* (Lindl.) Kuntze | Fagu | July-August | It is used in the preparation of Chyavanprash, to cure blood disorders, burning sensation in the body, male sterility, fever, dysentery, external and internal haemorrhage and general weakness. It is also used as an aphrodisiac, against insect bite and rheumatism. |
| 10 | *Liparis paradoxa* (Lindl.) Rchb. f. | Tara Devi | July-August | --------------- |
| 11 | *Calanthe tricarinata* Lindl. | Narkanda | April-May | It is used in the traditional system of medicine to cure jaundice, sores, eczema, and as an aphrodisiac . |

During the survey, 11 orchid species were reported from the selected places. According to the literature, most of the species are medicinally important. The selected areas are well known tourist's places. Every year, thousands of tourists visit these places. To attract the tourists and to provide quality facilities, construction work is going on with a very fast speed. Because of this, more and more natural habitats are being destroyed. Due to habitat destruction, urbanisation, illegal trade, absence of particular pollinators, poor regeneration and introduction of exotic species, natural populations of these orchids are decreasing with the alarming rate. As more and more tourists are visiting, they are using plastic bags, wrappers and plastic bottles, thus increasing the pollution in these areas. Grazing is also destroying the young seedlings, inflorescences and even pseudobulbs and tubers also. In orchids, requirement of a particular mycorrhiza is also responsible for low seed germination in nature. Population size of *Cypripedium cordigerum*, *Spiranthes sinensis*, *Neottia listeroids* and *Malaxis muscifera* is very small. These species require immediate attention. Pathak *et al.* ,(2005) have also observed the need of conservation of these orchids. Therefore there is immediate need to conserve these important plants using both *in situ* and *ex situ* methods. For the *in situ* conservation, plants can be conserved in their natural habitat. Local people should be made aware about the importance of these plants to them by organising seminars, workshops and discussions. Strict rules should be made for the tourists. They should not allow using non biodegradable objects like polythene bags, wrappers and plastic bottles. Grazing should be regulated. Farmers should use ecofriendly insecticides in their fields. For the *in situ* conservation, cost effective protocols should be developed to multiply these plants using plant tissue culture techniques where very small amount of material is required. Micro propagated plants should be hardened and transfer to field conditions. It will also reduce the collection pressure on natural habitats. In *in-vitro* micropropagation, the association of a mycorrhizal fungus is also not required, if supplied with suitable nutrients (Knudson, 1946). In this technique, different explants like shoot tip, leaf segments, inflorescence axis, flower buds, rhizome segments and immature seeds can be used. Immature seeds require less time period to develop into mature plants than other explants mentioned earlier.

## REFERENCES

Deva, S. and Naithani HB (1986) The Orchid Flora of North West Himalaya. Print and Media Associates, New Delhi, India.

Khan, MS., Page, S, Ahmad, H, Shaheen, H, Ullah, Z, Ahmad, M and Harper, DD. (2013). *Medicinal flora and ethnoecological knowledge in the Naran Valley, western Himalaya, Pakistan.* J. Ethnobiology and Ethnomedicine. Doi: 10.1186/1746-4269-9-4.

Knudson, L. 1946. *A new nutrient solution for germination of orchid seed. Am. Orchid Soc. Bull.* 15:214-217.

Hossain Musharof, Ravi Kant, Pham Than van, Budi Winarto, Songjum Zeng and J. Teixeira da Silva. 2013. *The Application of Biotechnology to Orchids.* Critical Reviews in Plant Sciences. 32(2): 69-139.

Medhi, RP., and Chakrabarti, S. (2009). *Traditional knowledge of NE people on conservation of wild orchids.* Ind. J. Traditional Knowledge. 8:11-16.

Mishra, AP and Saklani, S. (2012). *Satyrium nepalense: a rare medicinal orchid of western Himalaya (India); Phytochemical screening, antimicrobial evaluation and conservation studies.* Indonesian J. Pharm. 23(3): 167-170.

Pathak, P, Vij, SP and Verma, J. (2005). *Ground Orchids in North West Himalaya.* Proceedings of the 18th World Orchid Conferrence, Dijon, France, 11-12 March.

Vij, SP, Verma J, and Thakur K. 2011.. *Coeloglossum viride (L.) Hartman and Vanda cristata Lindl (Orchidaceae): Two new records for Himachal Pradesh.* Phytomorphology 61 (1, 2): 25-27.

Chapter 8

# Conservation of indigenous medicinal mushroom *Ganoderma lucidum* - An approach

**Uma Gowrie S., G. Chathurdevi and K. Rani**

Department of Plant Biology and Plant Biotechnology Ethiraj College for Women Chennai, Tamil Nadu

## ABSTRACT

*Ganoderma lucidum* commonly known as "The King of Mushroom" or "The Mushroom of Immortality" has been used for thousands of years by far Eastern Countries for medicinal properties. This mushroom is considered to be critically endangered in its natural habitat. A study of Biodiversity was carried out in our college campus (Ethiraj College for Women, Chennai, Tamil Nadu), and interestingly, the basidiocarp of *Ganoderma lucidum* was found on the dried wood logs of *Guazuma tomentosa* (Sterculiaceae). As a measure of conservation of the natural bioresource, a study on morphological variation, developmental stages of basidiocarp was carried out in *Ganoderma lucidum* with respect to environmental conditions. The microbial load analysis was carried out with the soil sample from the mushroom growing site to reveal the presence of fungal and bacterial population at different depths (surface, 5cm and 15cm) and at different zones (5cm and 30cm apart) to study the microbial interactions. Further antimicrobial potential of *Ganoderma lucidum* was tested using various concentrations of different extracts (chloroform, ethanol, methanol, ethyl acetate and water) of basidiocarp against the Bacteria (*Salmonella typhi, Bacillus subtilis, Escheriachia coli, Klebsiella pneumoniae, Pseudomonas aerunginosa*) and Fungi (*Aspergillus niger, Aspergillus flavus, Candida albicans, Curvularia species and Fusarium oxysporum*). The study revealed that the chloroform extract of basidiocarp of *G.lucidum* showed maximum antibacterial activity against *S.typhi* and antifungal activity against *C.albicans*. Preliminary phytochemical screening of *Ganoderma lucidum* was carried out for different extracts to detect the bioactive principles which can help in drug discovery and development. FT-IR analysis of the sample *G.lucidum* was also carried out to study the distribution of functional group within organic fractions. TLC and HPLC study confirmed the presence of triterpenoids and polysaccharides. Antioxidant activity of *Ganoderma lucidum* was also assayed.

## INTRODUCTION

The present world is desperately in search of new antibiotics because of alarmingly increase in bacterial and fungal resistance to existing antibiotics due to their inappropriate and indiscriminate use. Mushroom (wood decay fungi) is considered to have the antimicrobial and immense medicinal properties. (Hassan Iftekhar *et al.*, 2011) The response of microorganisms to mushroom extracts might vary depending upon the nature of environment in which it has been grown. (Iwalokum *et al.*, 2007)

*Ganoderma* species are regarded as higher fungi because the carpophores are visible enough to be seen with naked eyes. *Ganoderma lucidum* is commonly known as "The King of Mushroom" or "The Mushroom of Immortality" has been used for thousands of year's by far Eastern countries for its medicinal uses. It belongs to the family Polyporaceae, class Basidiomycetes. They are regarded as polypores because they possessed tiny pores underside their cap which contained reproductive spores. The caps are spongy when fresh, hardening to a shiny, smooth woody structure when matured. The color of the caps ranges from brown to yellowish with reddish-brown being typical of the polypore. The pore surface is cream in color and the spores are brown. *G.lucidum* is widely distributed all over in tropical and temperate regions. In natural forests, it occurs on old declining trees and decayed wood stumps causing white rots removing lignin as well as cellulose and related polysaccharides (Lalitha, 2004). This annual mushroom grows on a wide variety of dead or dying trees, e.g., deciduous trees especially oak, maple, elm, willow, sweet gum, magnolia, and locust (*Quercus, Acer, Alnus, Betula, Castanea, Coryolus, Fagus, Fraxinus, Populus, Pyrus, Magnolia, Tilia*) (Wasser *et al.*, 2005). *G.lucidum* is less frequently found on coniferous trees (e.g., Larix, Picea, Pinus) in Europe, Asia, and North and South America (in temperate rather than subtropical regions).

The fruiting bodies contains a variety of chemical substances, major components are terpenoids and polysaccharides, currently 130 triterpenoids and more than 100 types of polysaccharides are reported from *G.lucidum*. It has been reported that the major secondary metabolites of *G.lucidum* are ganoderic acid, triterpenes and polysaccharides. These bioactive compounds have been implicated for their high antioxidant, immune regulatory and hypoglycemic activities. *Ganoderma lucidum* has gained wide popularity as a health food in both China and Japan and it has many medicinal effects. (Lalitha., 2004). Although, *Ganoderma* species could not be consumed directly, they have been known all over the world as highly medicinal mushrooms (Jonathan and Awotona., 2010). But currently this mushroom is considered to be critically endangered in its natural habitat. The growth of this mushroom is greatly influenced by the environmental conditions, specific host species and the nature of habitat. A study of Biodiversity was carried out in our college campus (Ethiraj College for Women, Chennai, Tamil Nadu), and interestingly, the basidiocarp of *Ganoderma lucidum* was found on the dried wood logs of *Guazuma tomentosa* (Sterculiaceae). As an approach towards a measure of conservation, study on morphological variation, developmental stages of basidiocarp with respect to environmental conditions, their antimicrobial potential, phytochemical screening for different extracts of basidiocarp, FT-IR study, the presence of triterpenoid by TLC and HPLC, and anti oxidant assay were carried out in our present research work.

## MATERIALS AND METHODS

### Sample collection and identification

The mushroom species white rot fungi used in present study was collected from dried wood log of ***Guazuma tomentosa* (family-sterculiaceae)** at Ethiraj College for Women Campus, Chennai, Tamil Nadu and identified and authenticated by Dr.Kaviyarasan, Centre for Advanced Studies in Botany, University of Madras, Chennai-25.

### Meteorological studies

The environmental factors such as relative humidity, mean maximum and mean minimum temperature, rainfall, no. of rainy days, wind speed, light intensity during the period of growth for two consecutive years was collected for the location of Egmore from Regional Meteorological station, College road, Chennai-08.

### Morphological studies

The macroscopic characters, (shape, size, texture, pileus features, stipe, context, pore surface, tube layers) biomass - fresh and dry weight and the microscopic characters such as (hyphal system and basidiospore nature) and habitat characters (substrate or host, seasonality) were studied in order to classify this to the suitable species.

### Microbial analysis of soil sample

Soil samples were collected aseptically from different depths (surface, 5cm, and 15cm) and at different zones (5cm, 30cm apart) around the mushroom growing site. Nutrient agar medium and Potato Dextrose agar medium was used for bacterial and fungal growth respectively. Standard techniques were used to study the bacterial and fungal colonies from the soil samples.

### Crude extract preparation

The fresh fruit bodies of *G.lucidum* were dried and powdered. The powder was dissolved in the different organic solvents namely (chloroform, ethanol, methanol, ethyl acetate and water) and filtered after 48 hrs. These filtrates were concentrated at 40 to 50 degree centigrade using a flash evaporator (Fagade and Oyelade. 2009). The paste that was formed was freeze dried, and stored in a refrigerator in air tight containers for further study.

### Antimicrobial assay

Different extracts of the *Ganoderma lucidum* was assayed for its antibacterial and antifungal potential. Bacterial pathogens such as *Bacillus subtilis, Salmonella typhi, Escherichia coli, Klebsiella pneumoniae, Pseudomonas aerunginosa* and fungal pathogens such as *Aspergillus niger, Aspergillus flavus, Candida albicans, plant pathogens-Curvularia sps,* was used. Antimicrobial assay was tested using well diffusion method. Muller Hinton Media (bacterial media), PDA (Fungal media) were used. For anti bacterial assay, 24hrs old cultures were swabbed in the respective plates and wells were made on agar surface with sterile cork borer (7mm). For anti fungal assay, an aliquot (0.02ml) of spore suspension was used as inoculum for pour plate technique, after solidification, the appropriate wells were made on agar plates by using sterile cork borer. The different concentration of different extracts of basidiocarp was poured into well using sterile syringe. Gentamycin discs and Carbendazim were used as positive control. The plates were incubated at 37°C for bacteria & at room temperature for fungi. It was observed after 24 and 48hrs for zone of inhibition around the well.

## Phytochemical screening

Phytochemical screening for secondary metabolites for all the 5 extracts of basidiocarp was carried out by Mayer's test, Wagner's test & Hager's test (alkaloids), Fehling's test, Barfoed's test, Benedicts test (carbohydrates, glycosides), Million;s test, Biuret test, Ninhydrin test (protein and amino acid), spot test (fixed oil and fats), ferric chloride test, lead acetate test, alkaline reagent test (phenolic compounds and tannins), detection of gum and mucilages (Whistler and BeMiller, 1993), detection of terpenoids: (Salkowski test).(Harbone, 1976).

## Fourier Transform Infrared Spectrophotmeter (FT-IR)

For the FT-IR study, Spectrum FT-IR system (Shimadzu, IR Affinity 1, Japan), equipped with a DLATGS detector with a mirror speed of 2.8mm/sec. scan range: from 400-4000cm$^{-1}$ with a resolution of 4cm$^{-1}$ was used. The methanolic, ethanolic, chloroform extract of *G.lucidum* was prepared. The extract was evaporated by flash evaporator and it was mixed with a KBr salt, using a mortar and pestle and compressed into a thin pellet. Infrared spectra were recorded on KBr pellet on a Shimadzu FTIR spectrometer 4000 - 500cm$^{-1.}$ (Zeus *et al.*, 2013)

## Thin Layer Chromatography (TLC)

The solvent system used to identify triterpenoids was chloroform: methanol: water (30:4:1 v/v). The silica gel coated TLC plates were spotted with all 5 extracts of *G.lucidum* at equal distance. The plate was allowed to run until it reaches ¾ positions, 10% sulphuric acid in alcohol was used as spraying agent and plates were dried.( Nyoman *et al.*, 2001) The spots were noted under UV lamp

## High Performance Liquid Chromatography (HPLC)

2gms of powdered *G.lucidum* was mixed with 100ml of alcohol and kept for 24hrs in rotating shaker. Filtrate was collected and evaporated to dryness. A final volume of 1.5ml of extract was prepared as a stock solution with addition of methanol. HPLC equipment consisted of a SHIMADZU (Kyoto, Japan). LC 20 AD pump, an SPD 20 A- UV detector (SHIMADZU) with wavelength set at 254 nm and then LC solution software recorder (Integrator, Spectra-physics, Mountain view, CA). Room temperature was maintained. Separation was carried out at C18 column (Phenomenax 250) (Merck, Darmstadt, Germany). 20µl of the stock or sample solution was introduced with a Rheodyne valve equipped with 20µl external loop. The mobile phase rate was 1ml/min and consisted of Acetonitrile: Glacial acetic acid: HPLC grade distilled water (69.3: 0.01: 30.69) for triterpenoids, Distilled water (HPLC grade) for polysaccharides. Low pressure was maintained. (Ching-Hua Su *et al.*, 2001)

## Antioxidant Assay

((DPPH) RADICAL SCAVENGING ACTIVITY): 25mg of DPPH (150µm) was prepared in 100 ml of ethanol. To 1.9ml of DPPH, 0.1ml of extract of different concentration

(100,300,500,700&1000µg/ml) was added. Ethanol was used as blank solution. The absorbance of test mixture was read at 517nm. (Joseph *et al.*, 2009) The percent inhibition was calculated and expressed as percent scavenging of DPPH radical. The percent DPPH inhibition was calculated from the following formula:

$$\text{Percent DPPH activity} = \frac{\text{O.D of control- O.D of test} \times 10}{\text{O.D of control}}$$

## RESULTS AND DISCUSSION

**Morphological studies:**

The basidiocarp was collected from the tree host (plant) *Guazama tomentosa* at Ethiraj College for women, autonomous, Chennai-8 (August–October (2010-2012)).The result presented in table 1 shows the morphological characters of *G.lucidum* such as the existence of the stipe, shape of pileus, length or thickness of stipe, size or shape of pores, color of fruiting body, mycelial types and host range, microscopic characters and habitat characters. (Plate1). The morphological studies of basidiocarp of *G.lucidum* revealed that the isolate has similar character with that of *G.lucidum*.var.*lucidum*. (Gottlieb and Wright., 1999)

**Plate: 1 Growth stages of *Ganoderma lucidum***

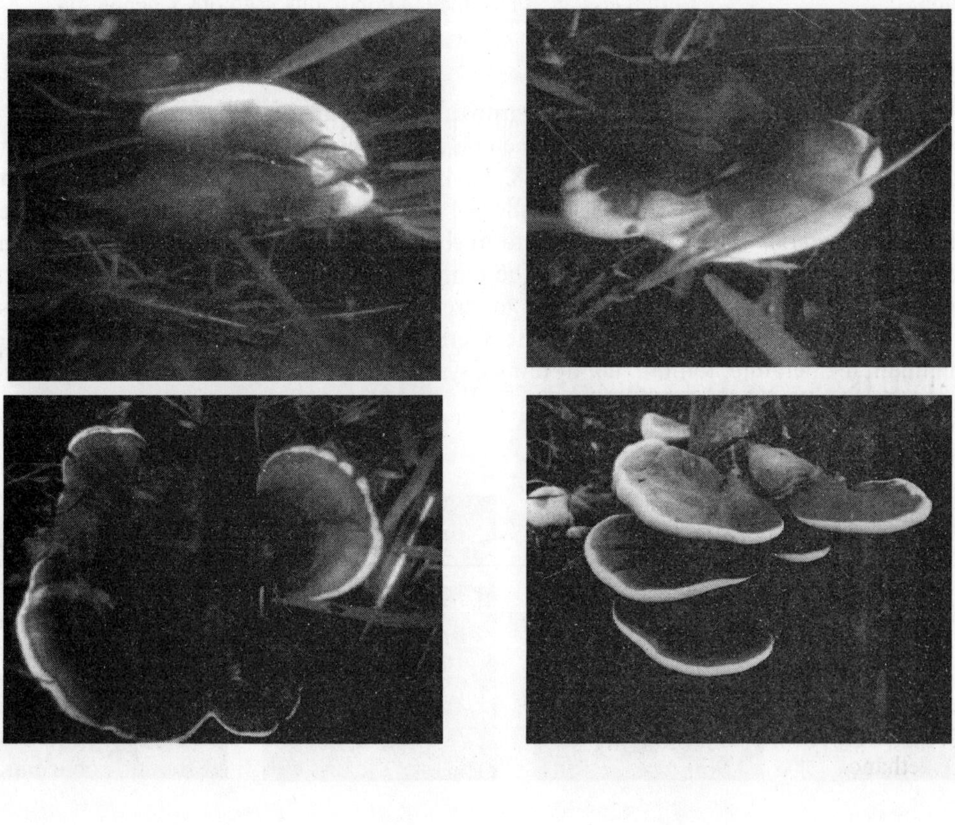

**Table: 1 Morphological studies**

| **Macroscopic characters** | |
|---|---|
| Shape | Applanate; sessile to centrally or laterally stipitate |
| Size | 17 to 19 cm length, 14 to 15cm breadth, 5to 6mm thickness |
| Texture | Tough, corky to nearly woody |
| Pileus | Dark reddish brown, varnish like crust, glaborous, often brown from spore deposit |
| Stipe | Eccentric/ lateral, upto 10cm in length, reddish brown, varnish like crust |
| Context | Creamy white; becoming dark purple with age, zonate, 3cm thick at base |
| Pore surface | Creamy white, becoming light buff, staining dark yellowish orange on bruising |
| Pores | Circular to angular, 4-5/mm |
| **Microscopic characters** | |
| Hyphal system | Dimitic, generative and skeletal hyphae |
| Basidiospores | Ellipsoid, 9-12×5.5-8µm |
| **Habitat characters** | |
| Substrate/host | fruiting found on dead trees and stumps of *G.tomentosa* |
| Seasonality | Annual |
| Type of decay | White rot |
| Location | Ethiraj college for women, autonomous, Egmore Chennai-08 |

## Morphological variations

Plate 2 shows morphological variations during growth phase of the basidiocarp of *G.lucidum*. The variations may be due to the environmental factors, prevailing in the sample collected location namely, Ethiraj College for women, Chennai-08. Environmental conditions also play a substantial role in the different morphological characteristics of *G.lucidum*. Elevated carbon-di-oxide level results in stem elongation in *G.lucidum*, other form shows "antlers" without a cap and these may be affected by carbon dioxide level as well. The morphological variations in *G.lucidum* such as formation of antilobe and fruiting bodies tend to stretch which is rare and was found to posses special medicinal properties. Table 2 shows the average relative humidity, mean maximum and mean minimum temperature, rainfall, no. of rainy days, wind speed, light intensity for continuous two years during its growth period.

**Plate: 2 Morphological Variations of *G.lucidum***

**Table: 2 Average Meteorological data during growh phase (Chennai, 2010 &2011)**

| Year | Mean maximum Temperature (°c) | Mean minimum Temperature (° c) | Mean relative humidity (%) | Monthly total rainfall (mm) | Number of rainy days (2.5mm &above) | Mean wind speed (KMPH) |
|---|---|---|---|---|---|---|
| 2010 | 33.3 | 25.7 | 80 | 183.8 | 9 | 4.3 |
| 2011 | 34.0 | 24.7 | 81 | 01.67 | 10 | 4.6 |

## Microbial load analysis

The microbial load analysis revealed the dominance of fungal population when compared with the bacterial population irrespective of different depths and different zones. It might be due to the presence of potent bioactive compounds found in basidiocarp of *G.lucidum,* which would have inhibited the bacterial population. Studies showing the antibacterial activity of *G.lucidum* crude extracts have proven the same in in-vitro studies of our work and also from Jonathan. *et al.*, 2010, where as the increased fungal population may be due to the mutual association existing among them. (Plate 3a&b)

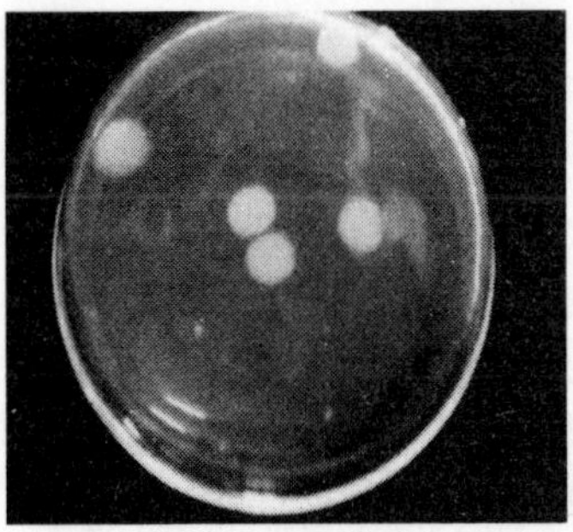

Plate:3a Bacterial colonies from the soil sample colonies from the soil sample

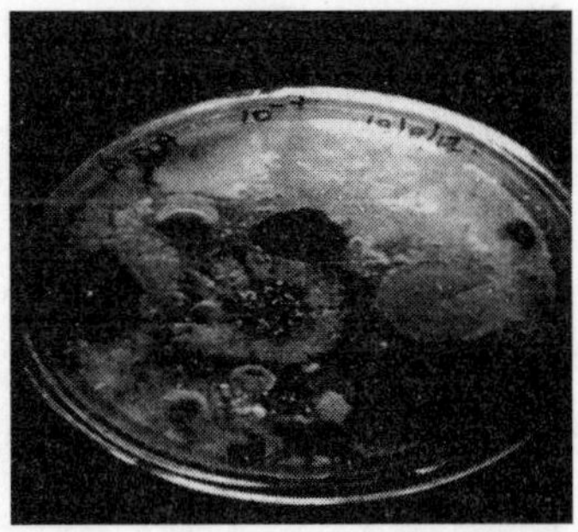

Plate:3b Fungal collected from the mushroom growing site

## Antimicrobial assay

The antimicrobial potential of *G.lucidum* done with various solvent extracts of different concentrations (25μg, 50μg, 100μg) revealed that chloroform extract of basidiocarp of *G.lucidum* (at 100μg) showed maximum antibacterial activity against *Salmonella typhi* (18mm) and anti fungal activity against *Candida albicans* (17mm).(Figure1a&b),(plate 4a&b). *G.lucidum* could be used for the invitro control of bacteria and fungi causing human diseases. It was observed that *G.lucidum*, extracts behaved differently in their antimicrobial effectiveness depending on the solvent used for extraction. (Sheena *et al*., 2003), the bioactive secondary metabolites of mushroom extracted may be different depending on the extractive solvents used. Water was observed in this study to be poor solvent compared with chloroform and methanol. Some phytochemical are more soluble in alcohol than in water. Difference may be due to its molecular configuration. The high level of effectiveness of difference as solvent of extraction could be linked with higher concentration of metabolites extracted. Cowan (1999) reported that the most active components are generally water insoluble, hence it is expected that low polarity organic solvents would yield more active extracts. The antibacterial activity of chloroform extract of *G.lucidum* from Iran, showed the growth inhibitory effects on *B.subtilis* & *S.aureus*. (Keypour *et al.*, 2008).

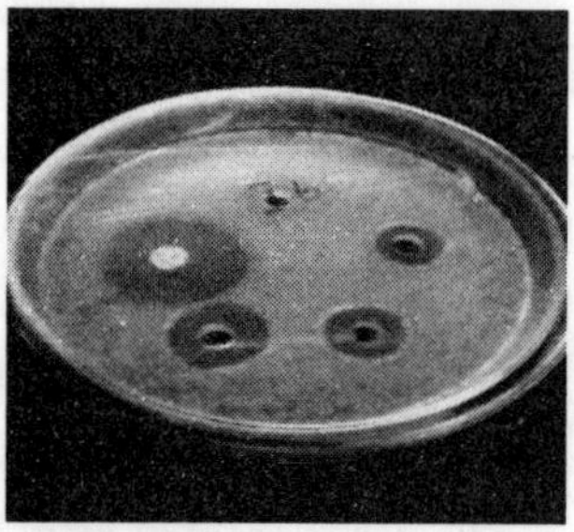

**PLATE: 4a Antibacterial activity of chloroform extracts of G.*lucidum* against *S.typhi.***

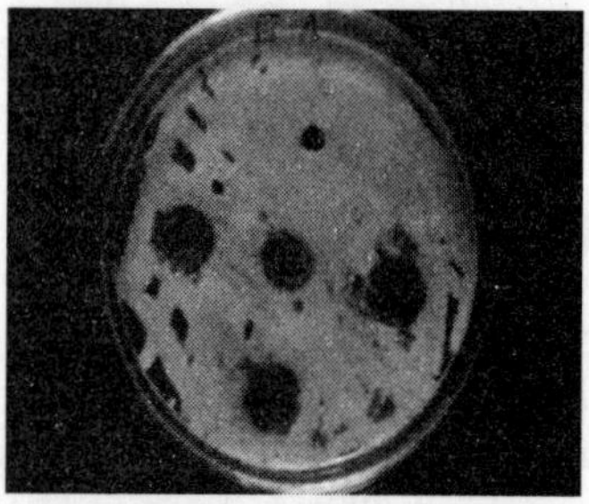

**Plate4b: Antifungal activity of chloroform extract of *G.lucidum* against *C.albicans.***

**Figure 1a: Antibacterial potential of different extracts of *G.lucidum***

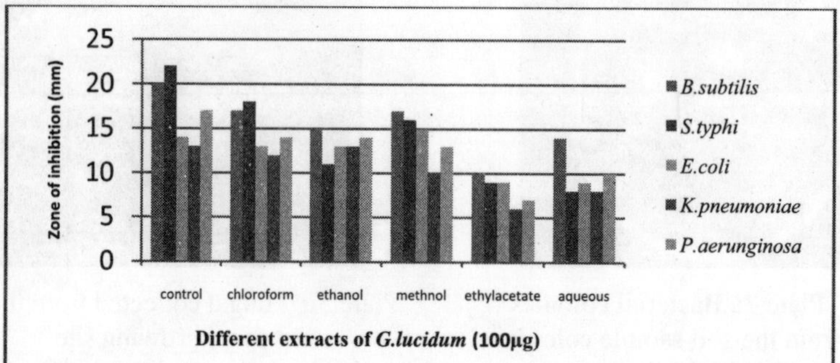


Control-gentamicin

## Phytochemical analysis

The phytochemical screening of *G.lucidum* for different extracts revealed the presence of alkaloids, carbohydrates, saponins, proteins, aminoacids, phytosterols, fixed oils and fats, triterpenoids, flavonoids, phenolic compounds and tannins in different tests. The polyphenols and flavonoids found in this study are known to be source of plant based antioxidants which can protect nerves, heart, liver and other organs and tissues (Table 3).The presence of alkaloids in the mushroom powder explains its antibacterial activity, since this phytochemical is reported to have antibacterial activity. (Iwalokun *et al.*, 2003). The presence of tannins which can complex with the metal ions and macro molecules such as proteins and carbohydrates obtained in powdered sample can be utilized in weight reduction management. Saponins have the anti-inflammatory, expectorant and immune stimulating effects (Joseph *et al.*, 2009).

**Figure 1b: Antifungal potential of different extracts of *G.lucidum***

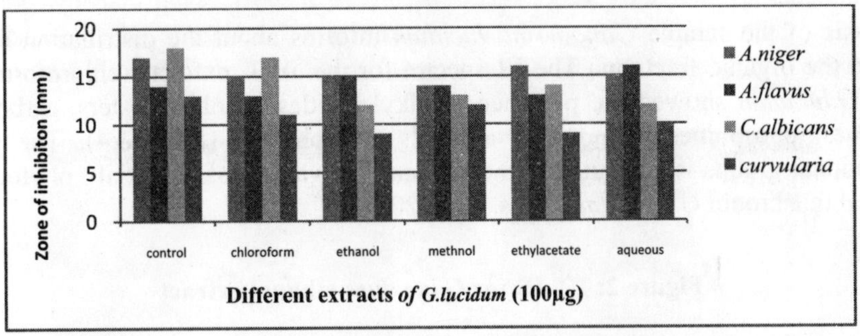


Control - Carbendazim

**Table: 3 Phytochemical screening of different extracts of *G.lucidum***

| S. No | Phytochemical Tests | Chloroform extract | Ethanol extract | Methanol Extract | Ethylacetate extract | Aqueous extract |
|---|---|---|---|---|---|---|
| 1 | **ALKALOID** | | | | | |
| | Mayer's test | + | + | + | + | + |
| | Wagner's test | + | - | - | + | - |
| | Hager's test | + | - | + | - | - |
| 2 | **CARBOHYDRATES &GLYCOSIDES** | | | | | |
| | Fehling's test | + | + | + | + | + |
| | Barfoed's test | + | + | + | + | + |
| | Benedict's test | - | - | + | + | + |
| 3 | **SAPONIN** | + | + | + | + | + |
| 4 | **PROTEIN AND AMINO ACID** | + | + | + | - | + |
| | Million's test | - | + | - | + | - |
| | Biuret test | - | - | - | - | - |
| | Ninhydrin test | | | | | |
| 5 | **PHYTOSTEROLS** | + | + | + | + | + |
| 6 | **FIXED OILS AND FATS** | | | | | |
| | Spot test | + | + | + | + | + |
| 7 | **PHENOLIC COMPOUNDS AND TANNINS** | | | | | |
| | Ferric chloride test | - | - | - | - | - |
| | Lead acetate test | + | - | + | + | - |
| | Alkaline reagent test | + | + | + | + | + |
| 8 | **TRITERPENOIDS** | + | + | + | ++ | + |
| 9 | **PHLOBATANINS** | - | - | - | - | - |
| 10 | **FLAVONOIDS** | | | | | |
| | Vanillin test | ± | ± | ± | = | ± |

## Fourier Transform Infrared Spectrophotmeter (FT-IR)

FT-IR analysis of the sample *Ganoderma lucidum* informs about the distribution of functional group within the organic fractions. The IR spectra for the all 3 extracts (chloroform, methanol, ethanol) of *G.lucidum* showed the presence of alkyl halides, alcohols, esters, carboxylic acids, ethers, alkenes, phosphines, monomeric alcohols and phenols (Fig2,3&4). The presence of various functional groups is attributed to the existence of variety of potential phytochemicals in this medicinal mushroom *G.lucidum.* (Zeus *et al.*, 2013).

**Figure 2: FT-IR for *G.lucidum* ethanol extract**

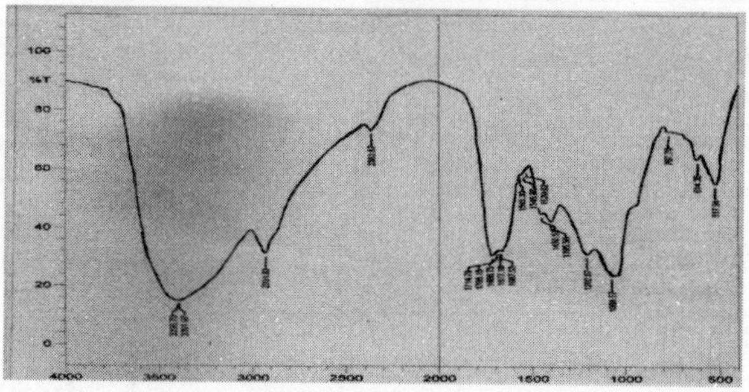

**Figure 3: FT-IR for *G.lucidum* chloroform extract**

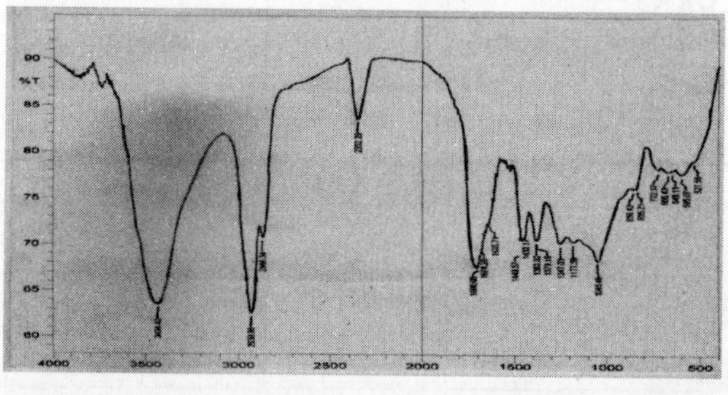

**Figure 4: FT-IR for *G.lucidum* methanol extract**

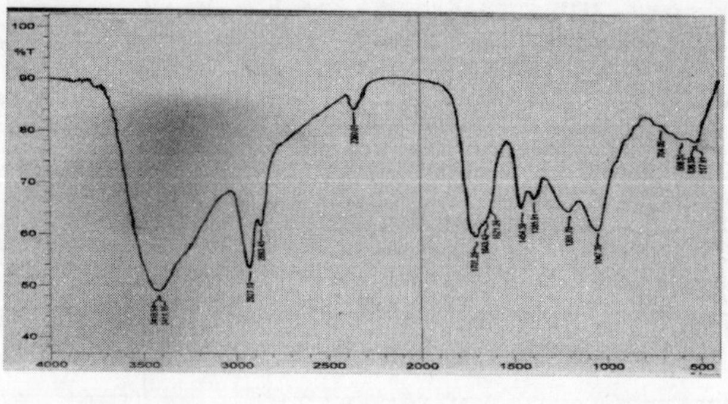

## Thin layer chromatography

The TLC study for triterpenoid was done for different extracts of *G.lucidum*. Plate with yellow to green fluorescent bands was observed under UV. It indicates the presence of triterpenoids. Strong spots were obtained for the chloroform and ethanol extracts of *G.lucidum* which has the $R_f$ value 0.271 and 0.312 respectively.(figure:5) Our result is in accordance with the result of Nyoman (2002), which showed the $R_f$ value ranging from 0.51 to 0.901 in benzene extract with carrprice reagent spray under UV light. Variation in $R_f$ value of triterpenoids may be due to the strains of *G.lucidum*, extraction solvents and reagents used in this study.

**Figure 5: TLC study for triterpenoids**

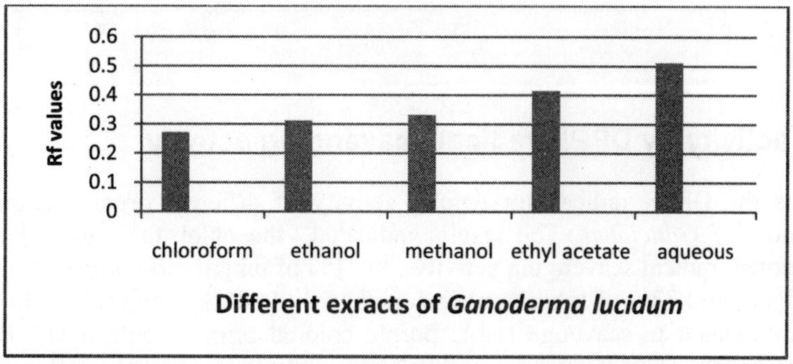


## High Performance Liquid Chromatography (HPLC)

HPLC study is carried out to identify and confirm the presence of triterpenoid and polysaccharides. Fig6 shows the chromatographic peaks 12.075 and 19.889 and confirms the presence of triterpenoids. The presence of polysaccharides in the samples of *G.lucidum* was also confirmed by HPLC (Fig 7) individual peak with the retention time of 5.056 with reference to Lalitha, (2004). Apart from this chromatographic peaks additional peaks are also present before, in between and after the chromatographic peaks of ganoderic acid C2(15.95) and B(21.93) which indicate the existence of additional triterpenoids (Ching-Hua Su, 2002). Triterpenoids of chloroform extract might be responsible for the antioxidant and inflammatory activities. These findings suggest that the therapeutic potentials of the chloroform extract of this mushroom is used for the prevention and the control of inflammation and diseases mediated through oxidative stress.

**Figure 6: HPLC study in *G.lucidum* for Triterpenoids**

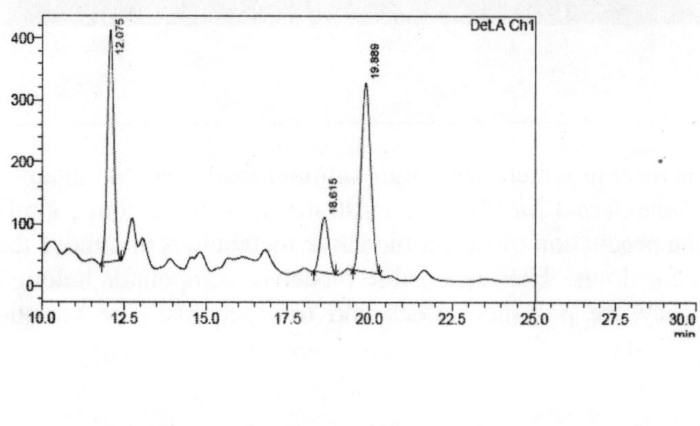

**Figure 7: HPLC study in *G.lucidum* for polysachharides.**

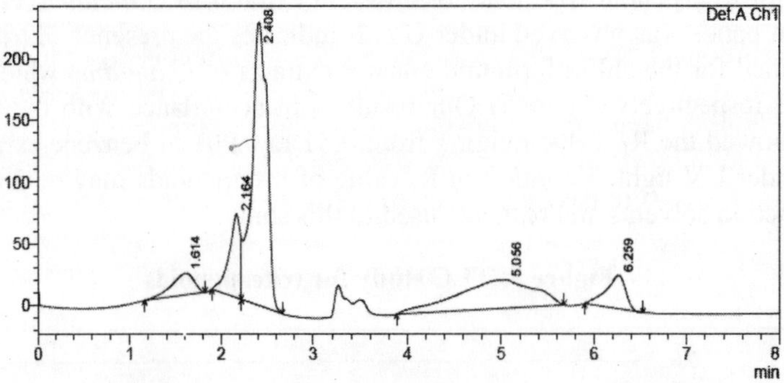


## Antioxidant activity by DPPH radical scavenging activity

Figure: 8 shows the DPPH radical scavenging activity of different organic extracts (chloroform, ethanol, methanol) of *G.lucidum*. The results indicated the chloroform extract at 1000µg/ml of *G.lucidum* has potent radical scavenging activity (89.71% of inhibition) compared to that of the other two extracts, ethanol (64.72% of inhibition) and methanol (74.05% of inhibition).In the DPPH assay, the ability of antioxidant to scavenge stable purple colored primary radical DPPH is tested by its depolarization at 515nm. This shows the ability of the extract to scavenge stable free radicals. The antioxidant property may be responsible for reduction of hepatic damage. (Joseph *et al.*, 2009).

**Figure 8: Free radical scavenging by DPPH assay of different extracts of *Ganoderma lucidum*.**

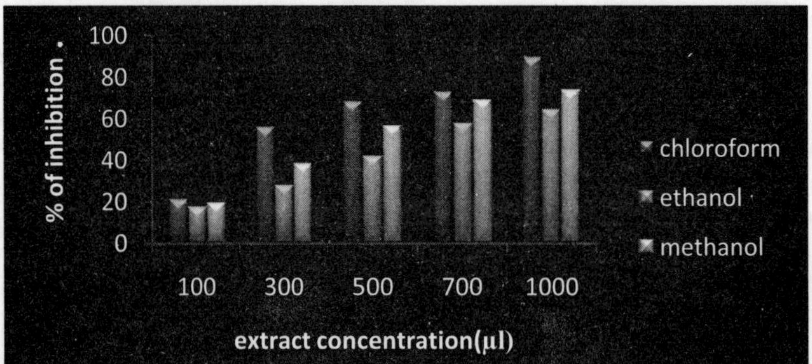


## CONCULSION

*G.lucidum* is the macrofungi which have high antimicrobial potential against various pathogens. Research work on *Ganoderma lucidum* demonstrates that mushrooms, similar to plants have a great potential for the production of useful bioactive metabolites and those they are considered to be prolific resource for drugs. The responsible bioactive compounds belong to several chemical groups very often they are polysaccharides and triterpenoids. The variation in the bioactive potentials of *Ganoderma lucidum* may be attributed to type of Host plant, soil, environmental condition existing in that area and geographical location. Hence, mushroom derived antimicrobial substances have received considerable attention in recent years. It is apparent from the present study that, the mushroom extracts from *G. lucidum* could be employed to combat several diseases caused by pathogenic microorganisms (Quereshi *et al.,* 2010). The conservation strategies that have been taken in our college campus for this indigenous mushroom are (1) Habitat is protected by planting more seedlings of host plants (2) Soil remains undisturbed (3) Periodical analysis of soil microbial load. A measure has been taken to cultivate *G.lucidum* in wood chips/ saw dust of *G.tomentosa*. Isolation and characterization of *G.lucidum* using appropriate media is carried out and the culture will be sent to MTCC for further identification and characterization to confirm the strain.

## ACKNOWLEDGEMENT

The author thanks the Head Dr.Mrs Shyamala Kanakarajan, the faculty members and non teaching staff of The Department of Plant Biology and Plant Biotechnology, Dr. Mrs Jyothi Kumaravel, Principal, Ethiraj College for women, Chennai for their valuable support and encouragement throughout the entire period of research.

## REFERENCES

Ching-Hua Su1, Yi-Zhen Yang1,Hsiu-O Ho, Chung-Hong Hu1, and Ming-Thau Sheu (2001). *High-Performance Liquid Chromatographic Analysis for the Characterization of Triterpenoids from G.lucidum* , Journal of Chromatographic Science, 39:93-100.

Cowan MM, (1999). *Plant Products as antimicrobial agents.* Clinical Microbiology Review, 12(4):564-582.

Fagade OE, Oyelade AA. (2009). *A Comparative study of the antibacterial activities of some wood-decay fungi to synthetic antibiotic discs*. EJEAF Che.: 8: 184-88

Gottlieb A. M, Saidman B. O, Wright J. E. (1998). *Isoenzymes of Ganoderma species from southern South* America. Mycol Res. ;102:415–26.

Hassan Iftekhar, Zubaida Khatoon Choudhry & Ismail Khan, (2011).*Comparative study of antibacterial activity of wood-decay fungi and antibiotics.* A Journal of the Bangladesh Pharmacological Society 6 (online).

Harbone, J.R. (1976). *Phytochemical methods; A guide to modern technique of plant analysis,* Springer, Berlin.

Iwalokun BA, Usen UA, Otunba AA, Olukoya DK.(2007).*Comparative phytochemical evaluation, antimicrobial and antioxidant properties of Pleurotus ostreatus.* African Journal of Biotech.; 6: 1732-39.

Jonathan S.G and Awotona F.E. (2010). *Studies on antimicrobial potentials of three Ganoderma species.* Afr.J.Biomed Res., 13. 133-139.

Joseph, S. Baby sabulal, George, V. Thozhuthumparambal P, Smina Kainoor, Janardhanan, K. (2009). *Anti-oxidative and Anti-inflammatory Activities of the Chloroform Extract of Ganoderma lucidum Found in South India.* Scientia pharmaceutica 77:111-121.

Keypour S, Riahi H, Moradali MF, Rafati H. (2008). *Investigation of the antibacterial activity of chloroform extract of Reishi mushroom Ganoderma lucidum from Iran.* International Journal of Medicinal Mushrooms, 10(4): 345-349

Lalitha.V, (2004.). *Extraction, Purification and Characterisation of extra polysaccharides from Ganoderma lucidum,* PhD thesis ,University of Madras, Chennai.

Nyoman P. Aryantha, Andi Adinda & Suswini Kusmaningati. (2001). *Occurrence of triterpenoids and polysaccharides on Ganoderma tropicum with Ganoderma lucidum as reference.* Journal of the Australasian Mycological Society, Inc., 20 (3) :23-29.

Quereshi S, Pandey AK, Sandhu SS, (2010). *Evaluation of antibacterial activity of different Ganoderma lucidum extracts.* Peoples J Scientific Res., 3: 9-13.

Sheena N, Azith TA, Thomas A, Mathew, Jonardhan KK. (2003). *Antibacterial activity of three macro fungi Ganoderma lucidum, Navesporus floccose and Phellium rimosus occuring in South India.* J Pharmaceutical Biol., 41: 564–67.

Wasser, P.S. Reishi or Ling Zhi (2005). *Ganoderma lucidum,* Institute of Evolution, University of Haifa, Mount Carmel, Isreal.

Zeus, S. Elumba, Franco G. Teves and Roberto M. Malaluan. (2013). *DNA-binding activity and in vivo cytotoxicity of Ganoderma applanatum (Pers.) Pat. supercritical-CO2 Extracts.* African Journal of Microbiology Research, 7(3): 202-210.

## Chapter 9

# Effect of Jasmonic acid and Salicylic acid on callus initiation and organogenesis in *Vigna mungo* (L.) Hepper using hypocotyl explant

**Asha A., S. P. Vairamathi and K. Lingakumar**

Centre for Research and Postgraduate Studies in Botany
Ayya Nadar Janaki Ammal College (Autonomous with CPE status) Sivakasi

### ABSTRACT

An attempt has been made to use the recent phytohormones such as Jasmonic acid (JA) and Salicylic acid (SA) as supplements in the growth medium to induce callusing in *Vigna mungo* using hypocotyl as explant source. Instead of testing the hormone individually, the hormones were combined with the conventional phytohormones to induce callus initiation. The callus response was high in MS medium supplemented with 0.5 ppm of 2,4-D and at 1 ppm of IAA. Likewise, Salicylic acid induced high percentage of callus induction at 1 ppm proving its efficiency in inducing the root formation. Similarly, Jasmonic acid induced high percentage of callus induction at 1.5 ppm but with greater rooting response than SA. Organogenic callus was observed at JA and SA supplementation. The combination of growth hormones such as NAA, BAP and SA was also tested for callus induction. The callus initiation response was high in MS medium supplemented with 0.75 & 1.5 ppm of NAA, 1 ppm of BAP and 0.5 ppm of SA (0.5 ppm). Thus, in future these hormones may serve to be efficient growth supplements for *in vitro* culture of agricultural crops.

## INTRODUCTION

Plant tissue culture is commonly used to describe the *in vitro* and aseptic growth of any plant part on a nutrient medium. Growth hormone is a natural chemical that exerts strong controlling effects on growth and development. It is used either in low or high concentrations to promote the callus or shoot and root formation. IAA and IBA are generally responsible for root induction whereas BAP and cytokinin are generally responsible for shoot induction and 2,4-D responsible for callus induction. Immature cotyledonary node explants produced a high frequency of plant regeneration in several crop species (Tivarekar and Eapen, 2001).

SA acts as a potential non-enzymatic as well as a plant growth regulator, which plays an important role in regulating a number of plant physiological processes (Fariduddin *et al.*, 2003; Singh and Usha, 2003; Waseem *et al.*, 2006; Arfan *et al.*, 2007). JA influenced a wide variety of physiological and developmental responses (Parthier *et al.*, 1992). The major functions of JA in regulating plant growth include growth inhibition, senescence and leaf abscission. JA has an important role in response to wounding of plants and systemic resistance for plants inhibiting the insect ability to digest protein.

## MATERIAL AND METHODS

### Explant sterilization and inoculation

*Vigna mungo* seeds were soaked in tap water for 30 minutes with few drops of Tween 20 and shaken well. The seeds were rinsed thrice in distilled water to remove the soap particles. The seeds were immersed in 70% alcohol for 30 seconds and further washed with distilled water. Seeds were further surface sterilized soaking in 0.1% (w/v) $HgCl_2$ for 4 minutes. The seeds were rinsed thrice in sterile water and germinated on double layers of pre-sterilized moistened filter paper. The seeds were subjected to UV-C (germicidal) radiation emitted from a UV-C lamp (Philips, India).

The seed samples were incubated for 16 hrs in a photoperiodic light simulation rack (Atlantis, India). The light intensity was 80 $W/m^2$ provided by cool white fluorescent lamps. The germinated hypocotyl was used as source of explant. Hypocotyl of explants 1 – 2 cm in length were excised using a sterile sharp blade in a Laminar air flow chamber (Atlantis, India) and implanted in MS medium supplemented with different concentrations of 2,4-D, IAA, Salicylic acid and Jasmonic acid.

### Callus induction and root regeneration

The hypocotyls of *Vigna mungo* were excised from *in vitro* germinated seeds and cultured on MS medium supplemented with 2,4-D (0.5 – 2.0 mg/L) and Salicylic acid (0.5 – 2.5 mg/L), IAA (0.5 – 1 mg/L), Jasmonic acid (0.5 – 2.5 mg/L), BAP + NAA (0.5 to 2 mg/L).

## RESULT AND DISCUSSION

**Callus initiation response in *Vigna***

The preparation of explants for inoculation was preceded by selecting healthy hypocotyls from 3–4 days germinated *Vigna mungo* seeds. The hypocotyls were horizontally sectioned using a sharp sterile blade to small bits (1–2 cm in length) and placed on callus initiation medium containing the growth hormones.

## Supplementation of 2,4-D in MS medium

The callus formation varied on MS medium supplemented with different growth regulators. Good callusing response was noticed at 0.5 ppm and 0.75 ppm of 2,4-D whereas moderate callusing response was noticed at 1.0 ppm and 1.5 ppm of 2,4-D (Fig. 3a). High concentration viz., 2 ppm showed least callusing response. The supplementation of 2,4-D in MS medium had resulted in induction of brown compact callus with hypocotyl explant (Table 1).

**Table 1. Effect of various concentrations of 2,4-D on callus induction from hypocotyl explants of *Vigna mungo* (L.) Hepper on MS medium (15 days after inoculation)**

| Concentration of 2,4-D (ppm) | Response of callus initiation | Frequency of callus induction (%) | Nature of the callus |
|---|---|---|---|
| 0.5 | +++ | 80 | Brown compact |
| 0.75 | +++ | 70 | Brown compact |
| 1.0 | ++ | 70 | Brown compact |
| 1.5 | + | 60 | Brown compact |
| 2.0 | + | 40 | Brown compact |

## Supplementation of IAA in MS medium

Regarding IAA, 0.5 ppm and 0.75 ppm showed high callusing response whereas meager amount of callus induction at 2 ppm of IAA was noticed (Fig.3b). High concentration such as 1.5 ppm of IAA showed a moderate callusing response by producing a brown compact callus (Table 2).

## Supplementation of SA in MS medium

The organogenic callus was induced from hypocotyls of *Vigna mungo* using SA. All concentration of SA showed positive response of callus induction (Table 3). But the frequency of callus was 40% at lower concentration of SA and maximum at 1 ppm of salicylic acid.

**Table 2. Effect of various concentrations of IAA on callus induction from hypocotyl explants of *Vigna mungo* (L.) Hepper on MS medium (15 days after inoculation).**

| Concentration of IAA (ppm) | Response of callus initiation | Frequency of callus induction (%) | Nature of the Callus |
|---|---|---|---|
| 0.5 | ++ | 70 | Brown compact |
| 0.75 | +++ | 60 | Brown compact |
| 1.0 | +++ | 80 | Brown compact |
| 1.5 | ++ | 50 | Brown compact |
| 2.0 | + | 30 | Brown compact |

All concentrations of SA produced well branched roots except that the nature of the root was different at each concentration. At 0.5 ppm of SA thin roots with slightly branched were produced whereas 1 ppm of SA produced well branched thick roots( Plate Fig.3c). At 1.5 ppm of SA branched roots were noticed. At high concentration of SA (2 ppm and 2.5 ppm), a low response was noticed. The effect of SA at various concentrations in inducing callus formation is shown in (Fig.1). Maximum response was noticed at 1 ppm.

**Table 3. Effect of various concentrations of Salicylic acid on callus induction from hypocotyl explants of *Vigna mungo* (L.) Hepper on MS medium (15 days after inoculation)**

| Concentration of Salicylic acid (ppm) | Response of callus initiation | Frequency of callus induction (%) | Nature of the callus |
|---|---|---|---|
| 0.5 | + | 40 | Organogenic |
| 1.0 | +++ | 70 | Organogenic |
| 1.5 | ++ | 60 | Organogenic |
| 2.0 | ++ | 50 | Organogenic |
| 2.5 | ++ | 40 | Organogenic |

SA has been reported to improve *in vitro* regeneration as well as abiotic stress tolerance in plants. The effect of various concentrations of SA or *in vitro* propagation of shoot apices of *Hibiscus* showed the regenerating potential of SA (Sakhanokho and Kelley, 2009). SA has been used to enhance *in vitro* regeneration in several other plants species (Quiroz-Figueroa *et al.*, 2001; Luo *et al.*, 2001; Hao *et al.*, 2006). In the present study, SA was found to induce root formation than shoot initiation. The response of root initiation was seen at all concentrations with maximum at 1 ppm. Addition of 0.5 mM SA induced shoot growth with more shoots per explant in *in vitro* culture of *Hibiscus* (Sakhanokho and Kelley, 2009).

## Supplementation of JA in MS medium

**Table 4. Effect of various concentrations of Jasmonic acid on callus induction from hypocotyl explants of *Vigna mungo* (L.) Hepper on MS medium (15 days after inoculation)**

| Concentration of JA (ppm) | Response of callus initiation | Frequency of callus induction (%) | Nature of the Callus |
|---|---|---|---|
| 0.5 | ++ | 60 | Organogenic |
| 1.0 | + | 50 | Organogenic |
| 1.5 | +++ | 100 | Organogenic |
| 2.0 | ++ | 70 | Organogenic |
| 2.5 | ++ | 60 | Organogenic |

+++ Represents maximum callus response , ++Represents moderate callus response, + Represents meagre callus response

Like SA and JA also produced organogenic callus. Both lower and higher concentration of JA induced roots. The frequency of root formation was 50% at 1 ppm of JA and maximum response (100%) at 1.5 ppm of JA was noticed (Table 4). All concentrations of JA produced roots. But the nature of the root was different at each concentration.

At 0.5 ppm, JA directly produced roots without undergoing callus formation. At 1.5 ppm of JA, long and well branched roots with 100% frequency of root formation was noticed (Fig.3d). At high concentration of JA (2 ppm and 2.5 ppm) moderate root formation like SA was observed. The supplementation of JA to the MS medium on callus induction in *Vigna* is shown in (Fig.1). Unlike SA, JA was found to induce highest response at 1.5 ppm. The effect of JA on shoot formation in cotyledons was examined by Patricia *et al.* (2001). Up to 1μM concentration, shoot formation was initiated whereas higher concentration (>10 μM) inhibited the number of shoots as well as shoot fresh and dry weight.

An initial exposure of *Pinus* cotyledons to JA resulted in shoot primordial and subsequent shoot formation (Patricia *et al.*, 2001). In contrast to Patricia *et al.* (2001), JA at 0.5 ppm to 2.5 ppm stimulated root primordial and subsequent root formation. Furthermore, extensive root formation with lateral branches was noticed under JA treatment. JA has been reported to influence shoot and bulb formation in Garlic at 1–10 μM concentration (Ravnikar, 1993).

**Fig 1. Concentrations of SA and JA in MS medium on the callus induction pattern in *Vigna mungo* (L.) Hepper**

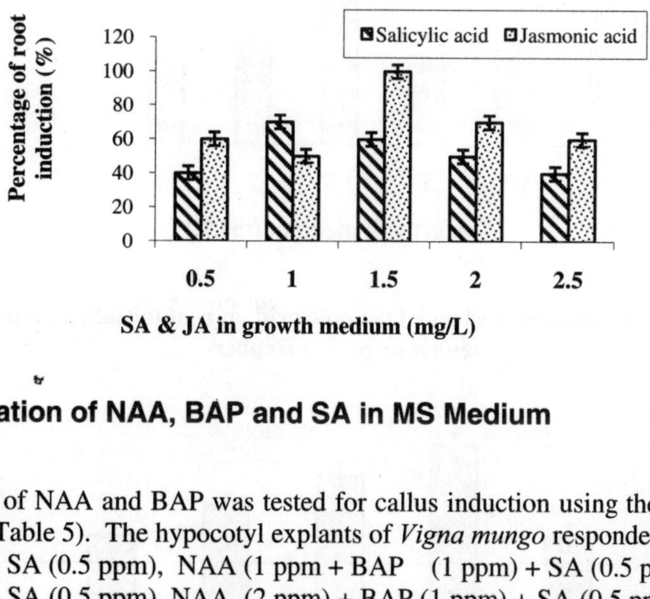


## Supplementation of NAA, BAP and SA in MS Medium

A combination of NAA and BAP was tested for callus induction using the hypocotyl explants of *Vigna mungo* (Table 5). The hypocotyl explants of *Vigna mungo* responded at NAA (0.75 ppm) + BAP (1 ppm) + SA (0.5 ppm), NAA (1 ppm + BAP (1 ppm) + SA (0.5 ppm), NAA (1.5 ppm) + BAP (1 ppm) + SA (0.5 ppm), NAA (2 ppm) + BAP (1 ppm) + SA (0.5 ppm) is shown in (Fig.2). The callus initiation response was high in MS medium supplemented with NAA (0.75 ppm and 1.5 ppm), BAP (1.0 ppm) and SA (0.5 ppm). Combined action of NAA (0.75 ppm) and BAP (1.0 ppm) on inoculated hypocotyl explants induced callus formation which was found to be friable in nature. Addition of SA (at 0.5 ppm) to this combination resulted in root initiation from the callus is shown in (Plate1 Figs a- e).

High frequency of callus induction was noticed from hypocotyl explant of *Vigna mungo* at 0.75 ppm of NAA with 1 ppm of BAP and 0.5 ppm of SA. Callus growth was highest (70%) when supplemented with IAA (3 mg/l) + BAP (2 mg/l) in *Boerhaavia diffusa* L. (Samiron Phukan *et al.*, 1999). In the present study, the addition of SA and JA to the MS medium resulted in enhanced callus initiation as well as root initiation. Compared to SA, JA was effective in inducing profuse rooting with lateral branching.

**Table 5. Effect of various concentrations of NAA, BAP & SA on callus induction from hypocotyl explants of *Vigna mungo* (L.) Hepper in MS medium (15 days after inoculation)**

| Concentration of NAA (ppm) | Concentration of BAP (ppm) | Concentration of SA (ppm) | Response of callus initiation | Frequency of callus induction (%) | Nature of callus |
|---|---|---|---|---|---|
| 0.75 | 1 | 0.5 | +++ | 80 | Organogenic |
| 1.0 | 1 | 0.5 | ++ | 60 | Organogenic |
| 1.5 | 1 | 0.5 | ++ | 70 | Organogenic |
| 2.0 | 1 | 0.5 | ++ | 60 | Organogenic |

**Fig. 2 The effect of various concentrations of NAA, BAP & SA on callus induction from hypocotyl explants of *Vigna mungo* (L.) Hepper in MS medium (15 days after inoculation)**

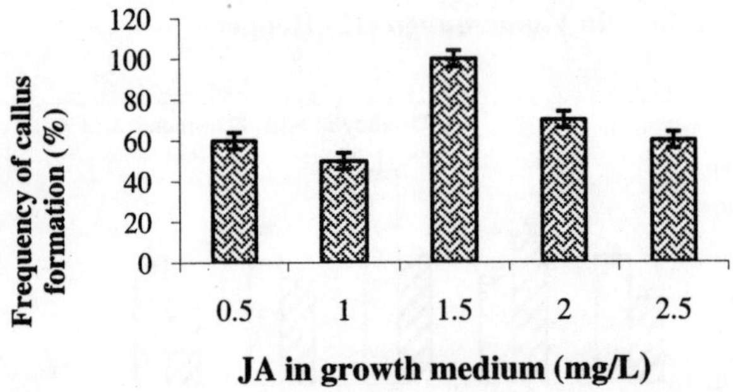


**Fig. 3 Effect of Salicylic acid and Jasmonic acid on root initiation response in *Vigna mungo* (L.) Hepper**

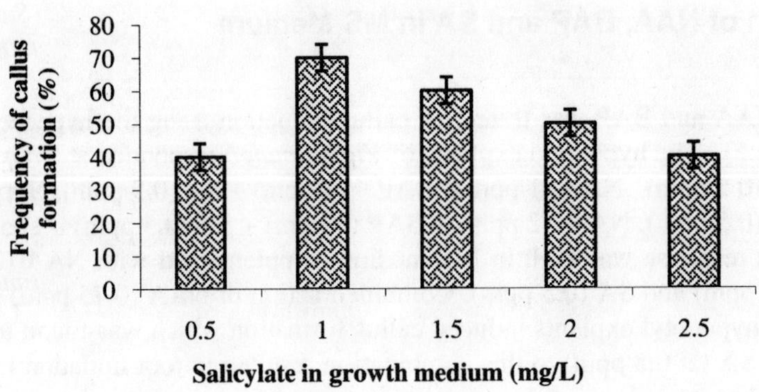

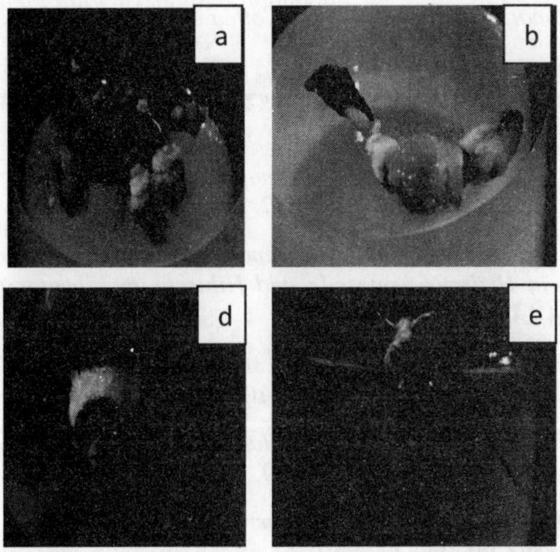


**Figs a-e) Induction of compact, organogenic callus and multiple roots from the hypocotyl explants of *Vigna mungo* (L) Hepper on MS medium supplemented with different concentration of 2.4-D, IAA,SA , BAP and JA.**

## REFERENCES

Arfan, M., H.R. Athar and M. Ashraf (2007). *Does exogenous application of Salicylic acid through the rooting medium modulate growth and photosynthetic capacity in two differently adapted spring wheat cultivars under salt stress?* J. Plant Physiol., 6(4): 685-694.

Fariduddin, Q., Hayat, S. and A. Ahmad, (2003). *Salicylic acid influences net photosynthetic rate, carboxylation efficiency, nitrate reductase activity and seed yield in Brassica juncea.* Photosynthetica 41: 281-284.

Hao, L., Zhou, L., Xu, X, Cao, J XiT, (2006). *The role of Salicylic acid and carrot embryogenic callus extracts in somatic embryogenesis of naked oak (Avena nuda).* Plant cell, Tissue Organ. Cult., 85: 109-113.

Luo-J-P, Jiang S-T, Pan L-J. (2001). *Enhanced somatic embryogenesis by Salicylic acid of Astragalus relationship with $H_2O_2$ production and $H_2O_2$ – adsurgens metabolizing enzyme activities.* Plant Sci., **161**: 125-132.

Parthier, B., C. Bruckner, W. Dathe, B. House and G. Herrmann. (1992). *Jasmonates: metabolism, biological activities and modes of action in Senescence and stress responses.* In: Karssen, C.M. and L.C. Van Loom, (eds.), Progress in Plant Growth Regulation. 88: 76-85. D. Vreugdenhil, Dordrecht: Kluwer Academic.

Patricia Tampe, A.. David M. Reid and Trevor A. Thorpe. (2001). *The effect of Jasmonic acid on de novo shoot formation in cotyledons of radiata pine (Pinus radiata D. Don).* J. Pl. Physiol. **5**: 607-611.

Quiroz-Figueroa, F. and M. Mendez-Zeel. (2001). *Picomolar concentration of Salicylates induce cellular growth and enhance somatic embryogenesis in Coffea Arabica tissue culture.* Plant Cell Rep., 20: 679-689.

Rao, M.V.,Koch, J.R., and K.R. Davis. (2000). *Ozone: a tool for probing programmed cell death in plants.* Plant Mol. Biol., 44: 345-358.

Ravnikar, M., Zel, J., Plaper, I. and A. Spacapan. (1993). *Jasmonic acid stimulates shoot and bulb formation of garlic in vitro.* J. Plant Growth Reg., 12: 73-77.

Sakhanokho, H. and M. Rowena Y. Kelley. (2009). *Influence of Salicylic acid on In vitro propagation and salt tolerance in Hisiscus acetosella and Hibiscus moscheutos (CV ' Luna Red'.* African J. Biotech., 8: 1474-1481.

Samiron Phukan, Uma Pathak and M.C. Kalita. (1999). *In vitro studies on Punnornova (Boerhaavia driffusa L.) – A medicinal herb.* Indian J. Plant Physiol., 4: 108-110.

Singh, B. and K. Usha. (2003*). Salicylic acid induced physiological and biochemical changes in wheat seedlings under water stress.* Plant Growth Regul. 39: 137-141.

Tivarekar, S. and S. Eapen.(2001). *High frequency plant regeneration from immature Cotyledon of mungbean.* Plant Cell Tissue Organ Cult., 66: 227-230.

Waseem, M., H.U.R. Athar and M. Ashraf. (2006). *Effect of Salicylic acid applied through rooting medium on drought tolerance of Wheat.* Pak. J. Bot., 38(4): 1127-1136.

## Chapter 10

# Effect of wild plant extracts against *Fusarium solani* and *Alternaria alternata* in *in vitro*.

**R. R. Saswade**

Department of Botany, Jijamata College of Science and Arts, Bhende
Tal-Newasa, Dist-Ahmednagar, Maharashtra, India
Email-saswaderrjcsna@rediffmail.com

**ABSTRACT**

Fungi are an important group of microorganisms responsible for the various diseases of plant species, causing a considerable loss in yield. A number of chemical fungicides are available in markets for the protection of plants. Some of them are excellent in terms of efficacy and cost benefit. However their indiscriminate use has created the problems of air, soil and water pollutions, development of resistance in target organisms and serious health hazards due to the toxicity of their residues. Efforts are being done for finding alternatives to chemical fungicides to overcome these problems. Plants like *Boswellia serrata, Ocimum americanum* and *Woodfordia fruticosa* leaf extracts of concentrations like 5%, 10% and 15% were used for testing their antifungal properties against the two important pathogens such as *Fusarium solani* (Mart.) Sacc and *Alternaria alternata* (Fr.) Keissler. The *Boswellia serrata* in all concentrations were found effective in antifungal activity as compare to *Ocimum americanum* and *Woodfordia fruticosa in vitro.*

## INTRODUCTION

Fungi are an important group of microorganisms responsible for the various diseases of plant species, causing a considerable loss in yield. A number of chemical fungicides are available in markets for the protection of plants. Some of them are excellent in terms of efficacy and cost benefit. However their indiscriminate use has created the problems of air, soil and water pollutions, development of resistance in target organisms and serious health hazards due to the toxicity of their residues. Efforts are being done for finding alternatives to chemical fungicides to overcome these problems. Plants like *Boswellia serrata, Ocimum americanum* and *Woodfordia fruticosa* leaf extracts of concentrations like 5%, 10% and 15% were used for testing their antifungal properties against the two important pathogens such as *Fusarium solani* (Mart.) Sacc and *Alternaria alternata* (Fr.) Keissler. Effective antifungal activity was observed in all concentrations of *Boswellia serrata* compared to *Ocimum americanum* and *Woodfordia fruticosa in vitro*.

## MATERIALS AND METHODS

Plants like *Boswellia serrata, Ocimum americanum* and *Woodfordia fruticosa* were collected identified from study area and used for testing their antifungal properties against the two important pathogens such as *Fusarium solani* and *Alternaria alternata*

*Fusarium solani* were isolated from wilted roots of chili and *Alternaria alternata* from spotted leaves of green gram. Materials cut into pieces and surface sterilized by 0.1%mercuric chloride and put on PDA medium for culturing at $28^0$C to $30^0$C.

Healthy fresh plant leaves will be select and washed thoroughly with fresh water and finally rinse with sterilized distilled water. 50 grams of plant leaves will cut into small pieces and mince with the help of grinder by adding 50 ml sterilized distilled water. These phyto-extracts will filtered through double-layered muslin cloth in 150 ml conical flasks and plugged with non-absorbent cotton. These filtered extracts will be autoclave at 1.2 kg $cm^{-2}$ pressure for 20 minutes and treated as a stock solution. The stock solution diluted as 5%, 19% and 15% concentration and it will use for testing antifungal property by well method.

## RESULTS AND DISCUSSION

**Table 1. Effect of various concentrations of plant extracts on the growth of *Fusarium solani in vitro*.**

| Name of the Plant Extracts | Concentrations of the extract | | | | | |
|---|---|---|---|---|---|---|
| | 5 % | | 10 % | | 15 % | |
| | Average colony diameter (mm) | Growth inhibition (%) | Averagr colony diameter (mm) | Growth inhibition (%) | Average Colony diameter (mm) | Growth inhibition (%) |
| *Ocimum americanum* | 82.74 | 5.90 | 78.10 | 13.10 | 72.00 | 19.70 |
| *Boswellia serrata* | 71.20 | 21.15 | 64.25 | 27.10 | 51.50 | 41.45 |
| Woodfordia fruticosa | 78.50 | 12.20 | 71.50 | 21.50 | 60.00 | 34.51 |
| Control | 89.50 | - | 89.00 | - | 89.50 | - |

**Table 2. Effect of various concentrations of plant extracts on the growth of *Alternaria alternata in vitro***

| Name of Plant Extracts | Concentrations of the extract | | | | | |
|---|---|---|---|---|---|---|
| | 5 % | | 10 % conc. | | 15 % conc. | |
| | Average colony diameter (mm) | Growth inhibition (%) | Average colony diameter (mm) | Growth inhibition (%) | Average colony diameter (mm) | Growth inhibition (%) |
| *Ocimum americanum* | 72.50 | 18.25 | 68.20 | 24.10 | 65.25 | 27.50 |
| *Boswellia serrata* | 70.25 | 21.90 | 58.50 | 34.15 | 51.00 | 42.80 |
| *Woodfordia fruticosa* | 74.75 | 15.50 | 67.25 | 25.10 | 62.00 | 29.90 |
| Control | 88.70 | - | 89.10 | - | 88.75 | - |

Anwar *et al.,* (1994) studied aqueous extracts of twentythree angiospermic plants for their antifungal properties against plant pathogens like *Alternaria alternata, Colletotrichum capsici, Curvularia lunata* and *Fusarium equisetti* and found that *Ocimum basilicum* and *Ocimum canum* to be effective against the pathogens treated. Chaudhary *et al.,* (1995) reported that oil extracted from *Ocimum canum* exhibited complete mycelial growth inhibition of *Fusarium monoliforme* causing of wilt of Sugarcane. Jain *et al.,* (1996) reported the antimicrobial activity of *Calotropis procera* against *Fusarium monoliforme, Curvularia lunata and Aspergillus flavus.* Misra *et al.,* (1997) showed that essential oil of *Callistemon lanceolatus* can be used to control *Fusarium monoliforme, Curvularia lunata and Aspergillus niger.*

Rahman *et al.,* (1997) found the antimicrobial activities of alkaloids of *Holarrhena antidysentrica* and *Crotolaria* against *Alternaria alternata, Fusarium equiseti* and *Macrophomina phaseolina.* Zaman *et al.,* (1997) reported neem and garlic extract to be effective against seed borne fungi of mustard *viz. Fusarium and Alternaria.* Deraniyagala *et al.,* (1998) showed the antifungal activity of *Ficus racemosa* leaf extract against *Fusarium* and *Alternaria.* Sharma *et al.,* (1998) found extracts of *Cassia italica* and *Cassia siamea* effective against *Fusarium monoliforme, Curvularia lunata, Aspergillus flavus* and *Aspergillus niger.*

Salahuddin *et al.,* (1998) reported antifungal activity of seed extract of *Aegle marmelos* against *Alternaria alternata, Fusarium equiseti* and *Curvularia lunata.* Purohit and Bohra (1999) reported seed coat leachates of Fenugreek, Cumin and Fennel in inhibition of germ tube growth in *Alternaria alternata, Fusarium oxysporum* and *Curvularia lunata.* Prakasam *et al.,* (2001) reported that *Azedirachta indica* can be used to manage post harvest diseases of carrot caused by *Fusarium solani.* Navi and Singh (2003) reported garlic extract to be effective in *Sorghum bicolor* mould caused by *Fusarium, Curvularia* and *Alternaria.* Rawal and Thakore (2003) reported the inhibitory effect of *Ocimum sanctum* on *Fusarium solani* causing Fusarium rot of sponge gourd (*Luffa cylindrica*). Verma and Dohroo (2003) reported that garlic extract was useful in inhibition of *Fusarium oxysporum* causing wilt of pea. Bhatnagar *et al.,* (2004) studied different plant extracts against *Fusarium oxysporum* and found *Datura* and *Plantago ovata* to be effective. Das and Bora (2004) proved the antifungal effect of *Capsicum annum* seed leachates on fungal spore germination of *Fusarium solani, Colletotrichum capsici, Alternaria alternata* and *Aspergillus niger*. Bohra and Mathur (2004) reported neem formulations effective in suppression of *Fusarium solani* causing root rot in soyabean. Sharma and Bohra (2003) showed the effect of certain medicinal plant extracts on *Fusarium oxysporum*. Zore *et al.,* (2004) studied the antifungal activity of two medicinal plants and found that *Glycirrhiza glabra* extract inhibited the growth of *Fusarium sp*:

## REFERENCES


Anwar, M.N.; Singha, P.; Begum, J. and Chowdhury, J. U., (1994). *Antifungal Activity Of Some Selected Plant Extracts On Phytopathogenic Fungi.* Bangladesh. Jour. of Life Sci. (Bangladesh). 6 (2): 23-26.

Bhatnagar, K.; Sharma, B. and Cheema, H. S., (2004). *Efficacy of Plant Extracts Against Fusarium oxysporum var cumini Wilt in Cumin* . Jour. Of Myco. Plant Patho. 34(2): 360-361.

Bohra, B. and Mathur, K., (2004*). Biocontrol Agents and Neem Formulations for Suppression of Fusarium solani, Root Rot in Soyabean.* Jour. Of Myco. Plant Pathol. 34(2): 408-409.

Chaudhary, R. K.; Rao, G. P. and Pandey, A. K., (1995). *Fungi toxicity of some essential oils against sugarcane pathogens.* .Jour. of Living World 2(2):30-37.

Das, R. and Bora, K. N., (2004). *Effect of Capsicum Annum Seed Leachates On Fungal Germination.* Jour. Of Myco. Plant Pathl. 34(2): 432-433.

Deraniyagala, S. A.; Wijesundera, R. L. and Weerasena, (1998). *Antifungal Activity Of Ficus racemosa Leaf Extract & Isolation Of The Active Compound.* Journal of the National Science Council of Sri Lanka. 26(1): 19-26.

Jain, S. C.; Sharma, R.; Jain, R. and Sharma, R. A., (1996). *Antimicrobial Activity off Calotropis procera,* Fitoterapia. 67(3): 275-277.

Navi, S. S. and Singh, S. D. (2003). *Effects Of Pounding & Garlic Extract On Sorghum Grain Mold and Grain Quality.* International Sorghum & Millets Newsletter.44: 122-124.

Prakasam, V; Abraham, S. and Kannan, C., ( 2001). *Management Of Post Harvest Fungal Diseases Of Carrot Using Botanicals.* South-Indian-Horticulture 49(Special): 271-274.

Rahman, M. S.; Anwar, M. N.; Begum, J. and Chowdhury, J. U., (1997). *Antimicrobial Activities Of The Alkaloids Of Three Plant Leaves.* Bangladesh Journal of Botany 26(1): 79-81.

Rawal, P. and Thakore, B. B. (2003). Investigation On Fusarium Rot Of Sponge Gourd Fruits. Jour. Of Myco. and Plant. Pathol. 33(1): 15-20.

Salahuddin Ahmed, Rahman, M. S.; Jasim, U. C.; Begum, J. and Anwar, M. N., (1998). *Antimicrobial Activities Of Seed Extracts and Crude Alkaloids Of Aegle Marmelos (L) Corr.* Chittagong University Journal of Science. 22(1): 77-81.

Sinha, R.P. and Sinha, B.B. ( 2004). *Effect Of Potash, Botanicals & Fungicides Against Wilt Disease Complex In Lenti.* Annals of Plant Protection Sciences 12 (2): 454-455.

Verma, S. and Dohroo, N. P. (2003). *Evaluation Of Botanicals In Vitro Against Fusarium Oxysporum Causing Wilt Of Pea.* Plant Disease Research Ludhiana. 18(2): 131-134.

Zaman, M. A.; Saleh, A. K.; Rahman, G. M. and Islam, M. T. (1997). *Seed-Borne Fungi Of Mustard and their Control With Indigenous Plant Extracts.* Bangladesh Journal of Plant Pathology. 13:1-2,25-28.

Zore, G. B.; Surwase, B. S. and Karuppayil, S. M (2004). *Antifungal Activity In Two Medicinal Plants*. Jour. Of Myco. Plant Pathol. 34:2, 543-545

Chapter 11

# Cellulase and xylanase enzyme activities in the basidiocarp of *Pleurotus spp*

**Padmaja C. K. and Shiji Wilson**

Department of Botany

Avinashilingam Institute for Home Science and Higher Education for Women (University)

Coimbatore, Tamilnadu

**ABSTRACT**

A laboratory investigation was carried out to assess the cellulase and xylanase enzyme activities in the basidiocarp (fruit body) of *Pleurotus ostreatus* and *P. cornucopiae.* The biochemical analysis of basidiocarp of *P.ostreatus* showed the presence of total carbohydrate (16.17 mg $g^{-1}$) and total protein content (0.133 mg $g^{-1}$) on par with *P. cornucopiae.* Among the cellulase enzyme complex, endo-β-1, 4 glucanase (1.56 $Uml^{-1}$ $mg^{-1}$ enzyme protein) was expressed at a higher level in *P. cornucopiae,* Exo- β-1, 4 glucanase and β –glucosidase activities were very much pronounced in *P.ostreatus* (0.87 and 1.29 $Uml^{-1}$ $mg^{-1}$) and xylanase activity (0.0583 $IUml^{-1}$ $mg^{-1}$ enzyme protein) was recorded maximally in the basidiocarp of *P. ostreatus* when compared to *P.cornucopiae.*

## INTRODUCTION

World-wide mushroom cultivation is dominated by the production of *Agaricus bisporus* (champignon), *Lentinula edodes* (shiitake), and *Pleurotus ostreatus* (oyster mushroom). The edible macro fungi are known as "Mushrooms", whereas the 'toxic/poisonous species are commonly referred to as "Toadstools. *Pleurotus* belongs to Basidiomycete class of fungi which has been regarded as an edible mushroom for many years. Many pharmaceutical substances with potent and unique properties were recently extracted from mushrooms and made their way all over the world. Mushrooms serve as delicacies for human consumption and as neutriceuticals "food that also cures". Mushrooms, the fruiting bodies of basidiomycetous fungi, contain substances of various kinds that are highly valued as medicines, flavouring and perfumes (Fan *et al.*, 2008).

Mushrooms are often called Queen of Vegetables and table delicious since time immemorial. Due to their nutritive qualities, they find place in recommended diets especially for heart patients and diabetics. Mushrooms are rich in proteins as compared to cereals, pulses, fruits and vegetables. Mushrooms also contain potassium and phosphorus. They contain low quantity of sodium, which make them ideal food for persons with heart and kidney problems. Since mushroom possess low caloric value, high protein, high fiber contents and high K: Na ratio, they are ideally suited for diabetics and hypertension patients. Mushrooms are nutritional functional food and as a source of physiologically beneficial and non toxic medicines. They have been reported to have significant pharmacological effects or physiological properties such as bioregulation, maintenance of homeostasis, and regulation of biorhythm, cure of various diseases, and prevention and improvement of life threatening diseases such as cancer, cerebral stroke and heart diseases. They have been also demonstrated to have effective substances for decreasing blood cholesterol, hypolipidemic, antithrombotic, hypotensive and other applications. The cultivation of the oyster mushroom *pleurotus spp* has increased greatly throughout the world during the last few decades (Mahendra *et al.*, 2005).

Cellulolytic enzymes play an important role in natural biodegradation process. Mushrooms secrete wide range of extra-cellular enzymes. These extra-cellular enzymes are attacking on the lignocellulosic material and convert complex organic matter into soluble substances, which can then, be absorbed by the mushrooms for the purpose of their growth and development of fruiting bodies. The present investigation aims to investigate the production of cellulose and xylanase activities in the basidiocarp of *Pleurotus ostreatus* and *P. cornucopiae.*

## MATERIALS AND METHODS

### Collection of materials

The basidiocarp of *Pleurotus ostreatus* and *P. cornucopiae* were collected from wooden logs in Palakkad district, Kerala State and were identified based on the morphology of the pileus and stipe of the basidiocarp.

### Preparation of the sample

The basidiocarp of the mushroom sample was shade dried and powdered. The powdered sample was used for the study.

### Preparation of enzyme assay from fungal basidiocarp of *pleurotus spp*

The basidiocarp of the fungus was washed twice with distilled water. A quantity of 1 g of the basidiocarp was macerated in 5 ml of Sodium acetate buffer (pH 5.0) for Exoglucanase Endoglucanase and filter paperaseactivity. For xylanase enzyme activity 5 ml of Sodium citrate buffer (pH 5.0) was used for enzyme assay. The homogenate was centrifuged in a refrigerated centrifuge (Plasto crafts - superspin R. Mumbai) at 10,000 x g for 15 minutes. The supernatant served as enzyme source.

## Biochemical analysis

- Estimation of carbohydrates (Anthrone Method - Hedge And Hofreiter., 1962)
- Estimation of Proteins (Lowry *et al.*, 1951)

## Enzyme activity

- Estimation of endo β–1, 4 Glucanase and exo– β –1, 4 Glucanase (Wang *et al.*, (1988).
- Estimation of Filter Paper activity (Total cellulase) (Gadgil *et al.*,1995)
- Estimation of Xylanase (Ec 3.2.1.8) (Baily,1988)

## RESULTS AND DISCUSSION

### Total carbohydrate (Table –I)

The total carbohydrate and protein content of the basidiocarp of *P.ostreatus and P. cornucopiae* were 16.17mg $g^{-1}$ tissue and 16.33 mg $g^{-1}$tissue and 0.133 mg $g^{-1}$ and 0.121 mgg $^{-1}$ tissue.

Similar view was expressed by Rashad *et al.* (2009) who obtained total carbohydrate content in the range from 20.88 to 33.00 % in the fruit bodies of *Pleurotus ostreatus* cultivated on papaya waste and citrus fruit waste.

The present finding is in agreement with the observation of Ponmurugan *et al.* (2004) who found that the protein content of fresh *Pleurotus ostreatus* basidiocarps were found to be 33.33 mg $g^{-1}$ and 28.38 mg $g^{-1}$ when grown in paddy straw and sugar cane molasses.

### Total carbohydrate and total protein content Table – I)

| S.No | Mushroom | Total carbohydrate (mg $g^{-1}$) | Total protein (mg $g^{-1}$) |
|---|---|---|---|
| 1. | *Pleurotus ostreatus* | 16.17 | 0.133 |
| 2. | *Pleurotus cornucopiae* | 16.33 | 0.121 |
| SED | | 0.04 | 0.002 |
| CD | (P<0.01) | 0.14 | 0.007 |

### Cellulase enzyme complex -Table – II)

Among the cellulase enzyme complex, the endo–β–1, 4 glucanase activities was more pronounced in *P.cornucopiae* (1.56 Uml $^{-1}$ mg$^{-1}$ enzyme protein) than *P.ostreatus (*1.28 Uml $^{-1}$ mg$^{-1}$ enzyme protein). The exo β 1, 4- glucanase activity in basidsiocarp of *P.ostreatus* was expressed maximally of (0.87 Uml $^{-1}$ mg$^{-1}$ enzyme protein) and minimally in *P.cornucopia* (0.38 Uml $^{-1}$ mg$^{-1}$ enzyme protein). The basidiocarp of *P.osteatus* registered highest FPase enzyme activity of (1.29 IUml $^{-1}$ mg$^{-1}$ enzyme protein) when compared to *P. cornucopia* (0.71 IUml $^{-1}$ mg$^{-1}$ enzyme protein).

Mishra (2009) reported the production of 0.144 $Uml^{-1}$ of endoglucanase, 0.137 $IUml^{-1}$ of filter paper activity and 0.041$Uml^{-1}$ of cellobiase enzyme activities in *Pleurotus florida* after 15 days of incubation with carboxymethylcellulose as carbon source.

Irfan *et al* (2010) investigated the production of cellulase activity of *Aspergillus niger* on pretreated sugarcane bagasse. The result of the study revealed that highest activities of CMC ase of 26.4 IU ml-1 and FP ase of 15.7 IU ml-1 were obtained in 1% Na2SO4 pretreated bagasse.

## Xylanase enzyme (Table – II)

The highest xylanase enzyme activity of 0.058 $IUml^{-1}$ mg-1 enzyme proteins was registered in the basidiocarp of *Pleurotus ostreatus* and 0.036 IU $ml^{-1}mg^{-1}$ enzyme protein by *P.cornucopiae.*

Kanmani *et al.* (2009) obtained increased amount of xylanase enzyme activity of 16.4 IU $ml^{-1}$ after 28 days of degradation of coir waste by *Pharnerochate chrysosporium* and *Rhizopus stolonifer*.

Similar finding was reported by Choudhary (2009). Who also obtained xylanase enzyme activity of 0.57 IU $ml^{-1}$ during the biodegradation of paddy straw by *Volcariella volvacea.*

The present result is in accordance with the findings of Irfan *et al* (2010) who also obtained highest xylanase activity of 60.31 IU $ml^{-1}$ with 5% $H_2O_2$ treated bagasse by ***Aspergillus** niger.*

**Table –II - Cellulase enzyme complex and xylanase enzyme activity**

| S. No | Mushroom | Cellulase enzyme complex *($U mg^{-1}$) | | | Xylanase ** ($U mg^{-1}$) |
|---|---|---|---|---|---|
| | | Endo–β–1, 4 glucanase | Exo–β–1, 4 glucanase | FP ase | |
| 1. | *Pleurotus ostreatus* | 1.28 | 0.87 | 1.29 | 0.0583 |
| 2. | *Pleurotus cornucopiae* | 1.56 | 0.38 | 0.71 | 0.0367 |
| SED | | 0.02 | 0.02 | 0.03 | 0.0023 |
| CD (P<0.01) | | 0.08 | 0.07 | 0.10 | 0.0073 |

* $U mg^{-1}$= one μ mol of glucose formed $min^{-1}$ $mg^{-1}$enzyme protein during hydrolysis reaction.
** $U mg^{-1}$= one μ mol of xylose formed $min^{-1}$ $mg^{-1}$ enzyme protein during hydrolysis reaction.

## CONCLUSION

Thus, it could be concluded from the present investigation that the enzyme profile of *Pleurotus ostreatus* and *P. cornucopiae* were found to be on par with each other. In order to maximize the usage of mushroom in our day to day life, we can preserve the mushroom in dried form. The utility of mushroom can be enhanced by preserving them in dried form and at the time of surplus production. So, the mushroom described as "Queen of vegetables and precious pearls of cookery" can be effectively harnessed for maximum cellulase and xylanase enzyme production which are of enormous industrial applications including pharmaceutical and textile industry.

## REFERENCE


Bailey, M. J. (1988). *A note on the use of dinitrosalicylic acid for determining theproducts of enzymatic reactions.* Appl. Microbiol. Biotechnol., 29: 494-496.

Choudhary , M., Dhanda S., Kapoor, S and Soni, G. (2009). *Lignocellulolytic enzyme activities and substrate degradation by Volvariella volvacea, the paddy straw mushroom/Chinese mushroom.* Indian. J. Agric .Res., 43 : 223 – 226.

Fan, L.,Soccol, R.C. and Pandey, A. (2008). *Current developments in SSf mushroom production,* Ch., 12, Springer, N.Y.,: 253-274.

Gadgil,N.J.,Daginawala. H.F., Chakrabarti.T and.Khanna, P. (1995). *Enhanced cellulose production by mutant of Trichoderma reesei.* Enzyme microbiol.technol, 17: 942-946.

Hedge, J.E. and Hofreiter, B.T., (1962). *Determination of total carbohydrates by anthrone method.* In: Carbohydrate Chemistry, Whistler, R.L. and Be Miller, J.N. (Eds.), Academic Press, New York; p. 163-201.

Hossain, S. Hashimoto, M., Choudhury E.K., Alam, N., Hussain, S., Hasan, M., Choudhury, S.K. and Mahmud, I. (2003). *Dietary mushroom (Pleurotus ostreatus) ameliorates atherogenic lipid in hypercholesterolaemic rats.* Clin. Exp. Pharmacol Physiol. 30: 470–475.

Irfan .M, Nadeem .M, Syed Q.A and Biag.S, (2010). *Submerged cultivation of Aspergillus niger on pretreated sugarcane bagasse* .World.J.Agricultural Sci. 6(4):466-472.

Kanmani, P., Karuppasamy, P., Pothiraj, C and Arul, V. (2009). *Studies on lignocelluloses biodegradation of coir waste in solid state fermentation using Phanerocheate chrysosporium and Rhizopus stolonifer.* African . J. Biotechnol. , 8: 6880 – 6887.

Lowry, O.H., Rosebrough, N.J., Farr, A.L. and. Randall, R.H.(1951). *Protein measurement with folin-phenol reagent.* J. Biol. Chem. 193: 265 - 275.

Mahendra R, Girish and Wasser, S.P. (2005). *Therapeutic potential of Mushrooms. Nat Prod Rad.,* 4:246-257. Mishra, B.K. (2009). *Cellulolytic enzyme activity of mushrooms grown on Carboxy methylcellulose.* Current Biotica., 3: 156 – 162.

Ponmurugan, P., Sekhar, S.J and Sreesakthi, T.R. (2007). *Effect of various substrates on the growth and quality of mushrooms.* Pak. J. Biol.Sci., 10: 171 – 173.

Rashad, M.M., Abdou, H.M. , Mahmoud, A. E. and . Nooman , M. U . (2009). *Nutritional Analysis and Enzyme Activities of Pleurotus Ostreatus Cultivated on Citrus Limonium and Carica Papaya Wastes* . Australian. J. Basic. Appl. Sci., 3: 3352-3360.

Wang, C, Hseu.T and Huang,C (1988). *Induction of cellulase by cello-oliosaccharides in Trichoderma konigii G- 39.* J .Biotechnol. 9:47-60.

Chapter 12

# Pathogenicity of root-knot nematode, *Meloidogyne incognita* on chilli plant (CV.NS-230)

**Bhavani S. and Shyamala Vijayavathi B.**

Department of Zoology, Avinashilingam Institute for Home Science and Higher Education for Women, Coimbatore – 641 043

**ABSTRACT**

A pathogenicity test was conducted against root-knot nematode, *Meloidogyne incognita* on chilli (*Capsicum annum*).The results showed that with the increase in the inoculum level of *Meloidogyne incognita* ie from 500 to 2000 nematodes /pot caused reduction in plant growth parameters viz., shoot leghth, root length, root weight, shoot weight and number of leaves. The highest reduction in plant growth parameters was recorded in the plants inoculated with the 2000 nematodes /pot. Number of galls and egg masses produced was also maximum at 2000 nematodes /inoculum levels.

## INTRODUCTION

Root-knot nematode (Meloidogyne spp) are most wide spread tiny organisms which limit words agricultural productivity (Sasser et al,1982:Taylor et al,1982).They attack almost all the cultivated plants but vegetables crops are their most preferred hosts (Sasser,1980). Among the plant parasitic nematodes, root-knot nematodes (*Meloidogyne* spp.) are economically the most important ones limiting agricultural productivity and quality (Javed *et al.*, 2006). They are sedentary endoparasites causing severe damage to a wide variety of economically important crops.Chilli (Capsicum annum) known to be a spice and vegetable crop ranked high amongst the economically important vegetables in the world. Their entry into the root system cause morphological symptoms such as stunting, yellowing, withering of leaves and root rotting.

## MATERIALS AND METHODS

Pathogenicity of root-knot nematodes (*Meloidogyne incognita*) on the host plant chilli (*Capsicum annum*) was studied by pot culture experiment using plastic tumblers under laboratory conditions at a temperature range of 30 – 40°C.

Four weeks old, two healthy seedlings were picked up and planted in each plastic tumbler pot (10 × 7 cm) containing 250 g steam sterilized soil. Five days after planting, second stage juveniles ($J_2$) collected from the egg masses were inoculated to the host plants (chilli) at three different inoculum levels i.e., 500 ($T_1$), 1000 ($T_2$) and 2000 ($T_3$) per pot around the root system. The roots were covered with sterilized soil followed by light irrigation. A control pot with plants without inoculation of nematodes was also set for comparison purpose. Each treatment was replicated thrice (Plate IIA). The experiment was terminated 35 days after inoculation. The plants were uprooted from the pots and the roots were washed gently to remove the adhering soil and dried with blotting paper. Data on plant growth parameters (shoot length, root length, shoot weight, root weight and number of leaves), number of galls and egg masses produced per root system were recorded and the data was analysed statically (ANOVA).

## Results

**Table I.Effect of *M. incognita* on the growth of chilli (cv. NS – 230) at different inoculum levels**

| Treatments | Shoot length (cm) | % decrease over control | Root length (cm) | % decrease over control | Shoot weight (g) | % decrease over control | Root weight | % decrease over control | Number of leaves | % decrease over control |
|---|---|---|---|---|---|---|---|---|---|---|
| Control (Uninoculated) | 14.97 | - | 5.57 | - | 2.78 | - | 0.77 | - | 10.34 | - |
| 500 nematodes / pot ($T_1$) | 12.30 | 17.8 | 4.08 | 26.8 | 2.28 | 17.9 | 0.50 | 35.0 | 8.50 | 17.8 |
| 1000 nematode / pot ($T_2$) | 10.30 | 31.2 | 2.97 | 47.0 | 1.51 | 45.6 | 0.36 | 53.3 | 6.00 | 41.9 |
| 2000 nematodes/ pot ($T_3$) | 7.45 | 50.2 | 1.60 | 71.3 | 0.50 | 82.0 | 0.24 | 68.8 | 4.00 | 61.3 |
| SED | 0.353 | | 0.389 | | 0.146 | | 0.030 | | 0.236 | |
| LSD 5% | 0.865 | | 0.951 | | 0.357 | | 0.074 | | 0.577 | |

Mean of three replications.

**Table II. Effect of different inoculums levels of *M. incognita* on gall and egg mass production in chilli plants**

| Treatments | Number of galls / root system | Number of egg masses / root system |
|---|---|---|
| Control (Uninoculated) | 0.00 | 0.00 |
| 500 nematodes / pot ($T_1$) | 4.00 | 2.50 |
| 1000 nematodes / pot ($T_2$) | 6.33 | 3.33 |
| 2000 nematodes / pot ($T_3$) | 12.33 | 8.0 |
| SED | 1.009 | 0.456 |
| LSD 5% | 2.468 | 1.116 |

Mean of three replications.

## CONCLUSION

Due to pathogenicity of *Meloidogyne incognita* the morphological changes observed in the host plants were stunted growth, yellowing and shedding of leaves in the above ground system and formation of root-knot or galls in the below ground system. These eco-friendly plants are suggested for the field application in large scale by the agro farmers.

## REFERENCES

Sasser, J.M. and Caster, C.C., (1982). *Overview of the international Meloidogyne project – Rational goals, implementation and progress to data plan.* Proceedings of the IMP Research and Planning Conference on Root-Knot Nematodes, *Meloidogyne* spp. (Region 3rd), Brasillia, Brazil, pp. 3-13.

Sasser, J.N.,(1980). *An advanced treatise on Meloidogyne* Vol. 1, Biology and control U.S.A., I.D.N.E. State University, USA, P. 422.

Sasser, J.M. and Caster, C.C., (1982). *Overview of the international Meloidogyne project – Rational goals, implementation and progress to data plan,* Proceedings of the IMP Research and Planning Conference on Root-Knot Nematodes, *Meloidogyne* spp. (Region 3rd), Brasillia, Brazil, pp. 3-13.

Javeed, N., El-Hassan, S., Gowen, S., Pemproke, B, Inam-Ul-Haq, M.,(2006). *The potential of combining pasteuria ponetion and neem (A. indica) formulation as a management system for root-knot nematodes on to Mato.* Eur. J. Plant Pathol. 120, 53-60

## Chapter 13

# Preliminary phytochemical and biochemical analysis of *Bacopa monniera*

**Gopika S. and K. Hemalatha**

Department of Botany
Avinashilingam University
Coimbatore

**ABSTRACT**

The present study evaluates the phytochemical and biochemical analysis of *B.monniera.* The leaf powder of *B.monniera* was dissolved in respective solvents *viz.,* petroleum ether, methanol,water and subjected to test the presence of secondary metabolites and biochemicals. Preliminary phytochemical analysis of petroleum ether leaf extract of *B.monniera* showed the presence of alkaloids, flavonoids, phenols, terpenoids, starch and cellulose .whereas tannins, quinones, steroids and fixed oils were absent. In the case of methanol extract almost all the above phytochemicals were present except starch and in the case of aqueous extract steroids, terpenoids and fixed oils were absent but all the other above phytochemicals were present. The biochemical analysis of leaf of *B.monniera* exhibited the protein content 4.72 mg /100g,the carbohydrate content 13.61 mg /100g and free amino acid 1.74 mg /100g respectively.

## INTRODUCTION

In ancient times, humans were searching for healing powers . But the world is witnessing a resurgence of interest in herbal drugs, cosmetics and food supplements. Thus, nature still serves as a primary source for the cure of various ailments. A wide variety of Plant derived compounds have been screened *viz.,* alkaloids, phenolic acids, flavonoids (Klimezak *et al.,* 2007), terpenoids, steroids, saponins, glycosides, lignins, tannins cartenoids and vitamins (Govindarajan *et al.,* 2005). They are distributed in different parts of the plant *viz.,* wood, bark, stem, leaves, root, pods, flowers, fruits and seeds (Mathew *et al.,* 2006 ; Chan *et al.,* 2005 and Cari *et al.,* 2004). The medicinal value of plants lie in secondary metabolities that produce a definite physiological action on body to prevent deterioration caused by reactive and harmful free radicals (Ahmed *et al.,*

2011).In the present study preliminary phytochemical and biochemical analysis of the leaf extract of *Bacopa monniera* have been investigated.

## MATERIALS AND METHOD

### Plant material

The plant sample *Bacopa monniera* (L.) Pennell was collected from Palakkad in the month of November. The leaves of *B.monniera* were shade dried for three weeks at ambient temperature in the laboratory under continuous ventilation away from sunlight and dust. The shade dried leaf was made to fine powder with homogenizer and the powder was sealed in polythene bags and stored away from light and moisture until the time of extraction.

### Preparation of leaf extracts

Different extracts were made from powdered leaf using petroleum ether, methanol and water. About 40 g of each leaf sample was soaked with the appropriate solvent. All the extracts were left under shaking conditions at room temperature for 24 hr. The resultant extracts were filtered using Whatman No. 1 filter paper and used for further analysis. All the analyses were done in triplicates.

### Biochemical studies

The leaf of *Bacopa monniera* was subjected to quantitative estimation of Proteins, Carbohydrate s and Free Amino acids using the following methods:

1 Estimation of Protein (Lowry *et al.*, 1951)
2 Estimation of Carbohydrate (Hedge and Hofreiter, 1962)
3 Estimation of Free Aminoacids (Moore and Stein, 1948)

### Preliminary phytochemical studies

The crude leaf extracts were subjected to qualitative tests for the detection of various primary and secondary plant metabolites such as alkaloids, terpenoids, flavanoids, steroids, phenol, cellulose, starch, tannin, fixed oil and quinone using the method described by Harborne (1984).

## RESULTS AND DISCUSSION

**Preliminary Phytochemical screening of the leaf extract of *B. monniera***

**Table 1 : Preliminary phytochemical analysis of leaf extract of *B. monniera***

| Plant name | Extract used | Alkaloids | Flavonoid | Tannin | Phenol | Steroids | Terpenoid | Quinone | Starch | Cellulose | Fixed oil |
|---|---|---|---|---|---|---|---|---|---|---|---|
| *B. monniera* | Petroleum ether | + | + | – | + | – | + | – | + | + | – |
| | Methanol | + | + | + | + | + | + | + | – | + | + |
| | Aqueous | + | + | + | + | – | – | + | + | + | – |

Preliminary phytochemical analysis of petroleum ether leaf extract of *B.monniera* showed the presence of alkaloids, flavonoids, phenols, terpenoids, starch and cellulose .whereas tannins, quinines, steroids and fixed oils were absent. In the case of methanol extract almost all the above phytochemicals were present except starch and in the case of aqueous extract steroids, terpenoids and fixed oils were absent but all the other above phytochemicals were present. A survey of literature reveals that the active principle of punarnava' is an alkaloid *B. verticillata* is also rich in alkaloid content (Goyal *et al.,* 2010)

## Biochemical analysis of the leaf of *B. monniera*

**Estimation of Protein, carbohydrates and free aminoacids**

The content of protein, carbohydrates and free aminoacid of the leaf of *B.monniera* is indicated in **Table 2 and Fig. 1.**

**Table 2 : Estimation of protein, carbohydrates and free amino acid content of *B. monniera***

| S. No. | Plant name | Protein (mg/100g) | Carbohydrates (mg/100 g) | Free amino acid (mg/100 g) |
|---|---|---|---|---|
| 1. | *B. monniera* | 4.72 ± 1.35 | 13.61 ± 2.31 | 1.74 ± 1.89 |
| SEd | | 0.71 | 1.72 | 2.08 |
| CD (0.05) | | 1.75 | 4.21 | 4.98 |

Values are expressed by Mean ± SD of three replicates.

The biochemical analysis of leaf of *B.monniera* revealed protein content as 4.72 mg /100g,the carbohydrate content 13.61 mg /100g and free amino acid 1.74 mg /100g respectively. Udayakumar *et al.* (2003) screened the protein content of the plant *Solanum xanthocarpum* through Million's reagent and estimated the protein content (3.9 mg/100 g). Naseerbanu *et al.* (2003) estimated carbohydrate content in *Amaranthus viridis* which showed higher carbohydrate content (35.62 mg/100 g). Gupta (1999) reported that the free aminoacid content decreased with increase in storage period of soyabean seeds. The results of quantitative estimation of primary metabolites showed higher amount of carbohydrates compared to protein and amino acid

**Figure 1: Estimation of protein, carbohydrates and free aminoacid content of *B. monneira***

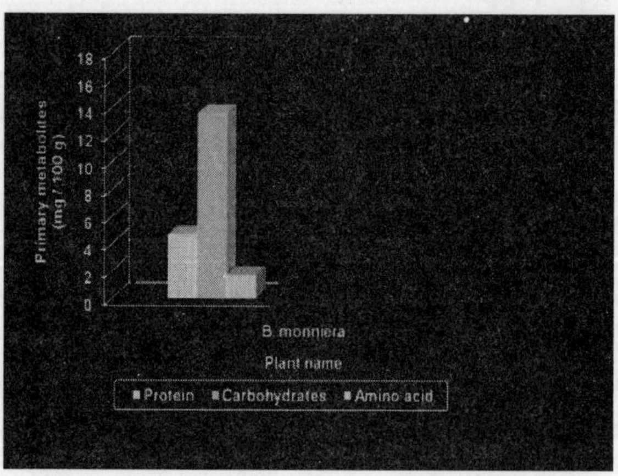

## CONCLUSION

In view of the large number of active principles produced by plants, we can only wonder at the incredibly vast reservoir, nature has to offer in which unlimited servers of medicinal ingredients are still largely untapped. The preliminary phytochemical and biochemical analysis of the leaf of *B.monniera* showed the presence and absence of different phytochemicals.

The presence of cellulose, phenols, tannins, flavonoids, quinone, steroids, starch, fixed oil, alkaloids, terpenoids were confirmed during phytochemical screening. Most of the phytochemical tests showed positive results with respect to methanolic leaf extract of *Bacova monniera.* But many of the phytochemicals were absent in petroleum ether and aqueous extracts of the leaf sample.Biochemical studies of the leaf powder of *B. monniera* showed the presence of protein, carbohydrates and amino acids with varied range.

## REFERENCES

Ahmed, H., A. Nustafa, Y. Osama and A. Thunibat, (2011). *Antioxidant activity of some Jordanian medicinal plants used traditionally for treatment of diabetes.* Pak. J. Biol. Sci., 11 : 351-358.

Chan, L.A., M. Mehta, A.R. Berenbaum, I. Zanger and N.J. Engeseth, (2000). *Honeys from different floral sources as inhibitors of enzymatic browning in fruit and vegetable homogenates.* Journal of Agri. Cultural Food Chemistry, 48 : 4997-5000

Cari, W.,(2004). *Medicinal plants as a source of antioxidants.* International Journal of Food Science and Technology, 39 (3): 2

Chakraborthy, G.S., (2009). Antioxidant activity of Abution indicum leaves, International Journal of Pharm. Tech. Research, 1 (4): 1314-1316. 95-3021-157.

Govindarajan, R., Vijayakumar, M. and P. Pushpangadan, (2005). *Antioxidant approach to disease management and the role of 'Rasayana' herbs of ayurveda.* Journal of Ethnopharmacology, 99 : 165-178

Goyal B.M., Bansal P., Gupta V., Kumar S., Singh R. and Maithani M. (2010). *Pharmacological Potential of Boerhaavia diffusa: An Overview.* International Journal of Pharmaceutical Sciences and Drug Research., 2(1), 18.

Gupta, A., (1999). *Studies on seed mycoflora, chemical and biological treatments on preservation and longevity of soya bean seeds.* Seed Tech. New 29 (4) : 63.

Harborne, J.B., (1984). Phytochemical methods. 11. Edn. In Chapman and Hall, New York, 21-26.

Hedge, J.E. and B.T. Hofreiter, (1962). Determination of total carbohydrate chemistry, Academic Press, New York, P. 17.

Klimezak, I., M. Nalecka, M. Szlachta and A. Glizezyns-ka-Swiglo (2007). *Effect of storage on the content of polyphenols, vitamin C and the antioxidant activity of orange juices.* J. Food Compos. Anal., 20 : 313-322.

Lowry, O.H., N.J. Rosenbourgh, A.L. Farr and R.F. Randall (1951). *Protein measurement with folin-phenol reagent.* J. Biol. Chem., 193 : 257-275.

Moore, S. and W.H. Stein, (1948). Polyphenol oxidase. In: Methods in enzymol . Eds. Colowick, S.P. and Kalpan, N.D.). Press, New York, 3: 468-470.

Naseer, B., A. Sivakumar and M.S. Subramanian, (2003). *Effect of manures on biomass production and pharmaco-biochemical properties of some green leafy vegetables*. Ancient Sci. of Life, 23 (2) : 123-134.

Udayakumar, R., H. Velmurugan, D. Sivanesan and R. Krishna,(2003). *Phytochemical and antimicrobial studies of extracts of Xanthocarpum.* Ancient Sci. Life 23 (2) : 90-96.

Chapter 14

# Effects of stressors on primary and secondary stress responses in *Cirrhinus mrigala* and its management using a herbal stress reliever stresroak

**Vasanthi J. and S. Binukumari**
PG and Research Department of Zoology
Kongunadu Arts and Science College
Coimbatore, Tamilnadu, India

**ABSTRACT**

Aquaculture has been the world's fastest growing food production system because of international demand for luxury items and regional need for food, high demand for fish products, declining wild stocks and decreased access to fisheries. Unfortunately, existing aquaculture operations can have a negative impact on the aquatic environment under conditions of stress, the body of the fish emits immediate responses recognized as primary and secondary responses. For the present study the effects of certain stressors on cortisol and blood glucose levels in *Cirrhinus mrigala* and the efficacy of a herbal stress reliever - stresroak was applied. Stresroak was tested in different concentrations. In the treatment groups of fishes showed a decreased cortisol and blood glucose levels. Stresroak showed a significant stress relieving effect at 30mg/l of immersion treatment. So Stresroak can be considered as a potent herbal stress reliever.

## INTRODUCTION

Stress is an unavoidable component in intensive aquaculture. Stress in fish can cause immune suppression and results in an increased susceptibility to diseases. Therefore, understanding of the physiology of fish and the careful management of the environment is required intensive aquaculture.

Herbs that relieve stress are becoming more common with wonderful result. Recently the herbal drugs attracted the attention of the fishery biologists, aquaculturists and scientists of animal husbandry (Unnikrishnan, 1995; Sambu and Jayaprakash, 2001; Ebanasar and Amutha, 2004; Ebanasar and Easwari, 2004). Herbal products need detailed evaluation and investigations before put into practical use.

The primary response is the perception of an altered state by the central nervous system(CNS) and the release of the stress hormones,cortisol and catecholamines(adrenaline and epinephrine) into the blood stream by the endocrine system.Secondary responses occur as a consequence of the released stress hormones,causing changes in the blood and tissue chemistry.e.g. an increase of plasma glucose.This entire metabolic pathway produces a burst of energy to prepare the fish for an emergency situation. The present work was conducted to determine the primary and secondary stress response in fish exposed to different stresses and to evaluate the efficacy of stresroak as a stress reliever in fish culture.

## MATERIALS AND METHODS

The fish *Cirrhinus mrigala* were procured from Aliyar fish farm at pollachi, Coimbatore, Tamilnadu. Fish were acclimatized to the laboratory condition for 10 days in glass tanks of 100 ltr capacity. Water was changed for everyday to avoid stress due to ammonia accumulation.The experimental animals were fed adlibitum with a balanced fish diet prepared in our laboratory.

Fishes were subjected to two types of stresses *i.e* salinity stress (10ppt for 35min) and crowding stress (30 fish were kept in 10 litre of water for 1hr) were applied to the fishes.Unstressed fishes were taken as the control and care was taken and it is free from stress.Blood samples was collected immediately after stress was given(0,6,24 and 48 hours) and the samples were tested for the primary and secondary stress responses such as cortisol and blood glucose.

Five groups of fish were maintained for testing the efficacy of stresroak as a stress reliever. Four groups of fishes such as T1, T2, T3 and T4 were exposed to the stresroak immersion in different concentration 10, 20, 30 and 40mg/l respectively for 24 hours. For the control stresroak was not applied after 24 hours and were again subjected to stresses and the blood samples were tested for the primary and secondary stress response at 0,6,24 and 48 hours). The data were expressed as means (SD).The test of significance between groups was determined by one factor ANOVA and groups were compared.

## RESULTS AND DISCUSSION

In the present study, the effect of different stressors in relation to cortisol and glucose level were estimated. Application of salinity stress elevates the cortisol level in stressed fishes were 182, 176, 153, 149mg/ml,fishes exposed to crowding stress showed the cortisol level of 92, 88, 43, 40mg/ml in 0, 6, 24 and 48 hours. It has been considered that elevated plasma cortisol level increases the susceptibility of fish to fungal bacterial and parasitic infection (Wedemayer, 1996).

Glucose level was 332, 216, 99 and 97 in salinity stress for 0, 6, 24 and 48 hours. In crowding stress the glucose production was 65, 63, 41, 41 respectively for 0,6,24 and 48 hours. Adrenaline and cortisol have shown to increase glucose production in fish and play an important role in the stress associated increase in plasma glucose concentration.

In the treatment of the experimental fishes showed the significant decrease in the cortisol and glucose level over the control fishes in all the stressed fishes.30mg/l of stresroak can relieve the stress in fish upto a normal level (Table1). The cortisol level was decreased upto 46, 52, 44, 40mg/ml for 0, 6, 24 and 48 hrs in salinity stress. At this concentration similar results have been observed in crowding stress. The same concentration of stresroak can decrease the glucose level from 22, 21, 18 and 13 mg/dl for 0, 6, 24, 48 hrs respectively and the same applicable for crowding stress.

The constituents of stresroak include *Phyllanthus emblica, Ocimum sanctum Wihania somnifera, Mangifera indica.* Each of these plants has been reported singly potent enough to boost immunity.

| Para meters | Time after Expo-sure (Hrs) | Control | Treatment Salinity Stress | | | | Control | Treatment Crowding Stress | | | |
|---|---|---|---|---|---|---|---|---|---|---|---|
| | | | T1 | T2 | T3 | T4 | | T1 | T2 | T3 | T4 |
| | 0 | 182±3.20 | 54±2.14 | 64±2.50 | 46±3.29 | 45±3.95 | 92±1.53 | 41±3.77 | 54±1.96 | 18±1.92 | 38±3.17 |
| Cortisol (mg/ml) | 6 | 172±2.88 | 61±2.16 | 51±3.64 | 52±3.29 | 48±2.71 | 88±2.86 | 33±1.88 | 53±2.33 | 16±2.53 | 43±2.49 |
| | 24 | 153±4.17 | 53±1.30 | 49±2.10 | 44±5.33 | 59±1.22 | 43±2.11 | 35±1.64 | 42±1.60 | 11±1.76 | 33±1.76 |
| | 48 | 149±1.19 | 52±3.23 | 48±1.83 | 40±1.78 | 47±1.14 | 40±1.77 | 30±1.19 | 41±1.39 | 11±1.33 | 26±1.53 |
| | 0 | 332±2.14 | 32±2.72 | 64±2.81 | 22±3.33 | 76±4.91 | 65±1.92 | 172±3.16 | 99±1.96 | 52±3.96 | 159±2.84 |
| Glucose (mg/dl) | 6 | 216±2.81 | 26±2.72 | 61±8.60 | 21±4.19 | 39±5.13 | 63±2.12 | 159±2.94 | 46±1.83 | 53±3.82 | 161±2.81 |
| | 24 | 99±2.15 | 23±1.33 | 42±5.48 | 18±1.72 | 34±2.12 | 41±1.29 | 152±2.16 | 107±0.09 | 52±2.54 | 172±1.63 |
| | 48 | 97±2.09 | 20±1.29 | 34±2.39 | 13±2.14 | 30±1.97 | 41±0.07 | 134±1.82 | 129±0.07 | 64±2.29 | 186±1.63 |

## REFERENCES


Unnikrishnan, G. (1995).*Effect of livol on growth food utilization and body composition of the Indian major carp Catla catla(Ham).* M.Sc.dissertation, University of Kerala, pp.34.

Sambu.C and Jayaprakash, V., (2001). Livol(IHF-1000) a new herbal growth promoter in white prawn, Penaeus indicus(crustaceae). Indian J.Mar.Sci.,30:38-43.

Ebanasar, J and .Amutha, M (2004).*Effect of dietary supplement of superliv on digestive enzyme of Cirrhinus mrigala* paper presented in National Symposium on Recent trends in Environment and Biotechnology,Government college(Autonomous).Kumbakonam.

Ebanasar,J and.Easwari, M. (2004). Influence of dietary supplement of superliv on morphology of liver in Cirrhinus mrigala paper presented in National Symposium on Recent trends in Environment and Biotechnology, Government college(Autonomous). Kumbakonam.

Wedmeyer, G.A.,(1996). *Physiology of fish intensive Culture Systems.* Chapman and Hall.115 Fifth Avenue New York, p. 232.

# Chapter 15

# Antibacterial activity of *Centella asiatica (L.)*

**T. V. Binu and C. D. Varghese**

Department of Botany

St. Thomas' College, Thrissur, Kerala, India

**ABSTRACT**

Plant based remedies have always been an integral part of traditional medicine throughout the world. The increasing demand for herbal medicines in the recent years is due to fewer side effects in comparison to synthetic drugs and antibiotics. A government program was launched in 1993 for the implementation by an NGO called "Foundation for Revitalization of Local Health Traditions". The program was aimed at creating awareness about the medicinal plants. Because of the increasing popularity of medicinal plants a study was conducted to analyze the antibacterial activity of *Centella asiatica (L.)*. *Centella asiatica (L.)* coming under the family Apiaceae (Umbelliferae) is prostrate, faintly aromatic stoloniferous perennial herb, with a glabrous stem and long petiolated fleshy leaves rooting at the nodes. The flowers are greenish to pinkish-white and are borne in dense umbels. Both alcoholic and water extracts of the whole plant were used for the study. The pure cultures of human pathogen, *Klebsiella pneumoniae Escherichia coli* were isolated from the clinical samples. .The antibacterial activity was screened by nutrient agar medium with disc diffusion method .Gatifloxacin, an antibiotic used to compare the antibacterial activity. The antibacterial activity of the plant extracts ere determined by measuring the diameter of inhibition zone of plant extracts and the control solution. Of the two extracts ethanolic extract showed maximum result.

## INTRODUCTION

Infectious disease is the world's leading cause of premature deaths, killing almost 50000 people every day. Morbidity and mortality due to diarrhea continues to be a major problem in many developing countries, especially amongst children. Infections due to variety of bacterial etiologic agents such as pathogenic *Escherichia coli, Salmonella spp., and Staphylococcus aureus* are most common. In recent years drug resistance to human pathogenic bacteria has been commonly reported from all over the world .With the continuous use of antibiotics microorganisms have become resistant. In addition to this problem, antibiotics are some time associated with adverse effect on host which includes hypersensitivity and allergic reactions. This has created immense clinical problems in the treatment of infectious diseases Alavijeh et al. (2012). Therefore, there is a need to develop alternative antimicrobial drugs for the treatment of infectious disease s; on approach is to screen local medicinal plants for possible antimicrobial properties. Though much is known about the chemistry and the antimicrobial action of several phytochemicals, very few reports are available on the possible mechanism of action. This report is the antibacterial activity of *Centella asiatica (L.)* against two human pathogens, *Escherichia coli* and *Klebsiella pneumoniae.*

## MATERIALS AND METHODS

The whole plant of *Centella asiatica* (L.) was collected from the college campus itself. The material is shade dried powdred and sieved. The pure cultures of human pathogens, *Escherichia coli* and *Klebsiella pneumoniae were* isolated from the clinical samples. 5gm of powder is taken and added 70% of ethyl alcohol, shaken continuously and kept for overnight. The extract was taken and residue discarded. Like that its water extract also taken. The nutrient agar medium was prepared, sterilized and poured into sterile petriplates and allowed to solidify.

The culture suspension was swabbed on the nutrient agar medium and incubated for 24h at 37 ºC. Three filter paper discs, made of Whattmanns' No; 1 filter paper, were placed on a single petridish .Two of them namely A&B were soaked with the plant extracts. The third disc namely G soaked with the antibiotic solution (Gatifloxacin).

The antibacterial activity of the plant extracts were determined by measuring the diameter of zone of inhibition of plant extracts and the control solution. The result was the compared with the zone of inhibition of the antibiotic solution.

## RESULTS AND DISCUSSION

The zone of inhibition was measured for both extracts and antibiotic solution and the Sensitivity pattern of *Centella asiatica (L.)* against E.Coli results depicted in Table 1 .

**Table 1 Sensitivity pattern of *Centella asiatica (L.)* against *E. coli***

| Samples | Inhibition zone (mm) | | | | | | Average zone diameter(mm) |
|---|---|---|---|---|---|---|---|
| A | 27 | 28 | 30 | 29 | 31 | 31 | 29.3 |
| B | 20 | 18 | 20 | 19 | 20 | 21 | 16.3 |
| G | 36 | 38 | 39 | 39 | 40 | 40 | 38.3 |

A –Dried ethanolic extract, B-Dried water extract, G Antibiotic –Gatifloxacin.

Here the maximum result was shown by dried ethanolic extract as compared to water extract. Similar result was shown by ethanolic extract of *Punica granatum* against Escherichia coli Voravuthikunchai et al. (2004).

Ushimaru et al. (2012) evaluated the interactions between crude extracts from *Psidium guajava, Zingiber officinale* and *Allium sativum* samples against *Escherichia coli* strains. The results showed that the potential use of these medicinal plants against *E. coli* strains is very useful.

**Table 2 Sensitivity pattern of *Centella asiatica (L.)* against *Klebsiella pneumoniae***

| Samples | Inhibition zone (mm) | | | | | | Average zone (mm) |
|---|---|---|---|---|---|---|---|
| A | 22 | 22 | 23 | 24 | 24 | 24 | 23.1 |
| B | 15 | 16 | 18 | 17 | 18 | 19 | 17.1 |
| G | 25 | 24 | 26 | 25 | 26 | 28 | 25.6 |

Table 2 shows that, comparing to water extract and antibiotic, dried alcoholic extract showed the maximum result. The average of zone of inhibition is 23.1. Sharma et al. (2009) reported that ethanol extract of *Azadirachta indica* and *Ocimum sanctum* exhibited antibacterial activity against *Klebsiella pneumoniae.*

## CONCLUSION

The results have revealed the medicinal potential of the plant in the treatment of diseases. Dried alcoholic extract showed a promising support on human disease therapy.

## REFERENCES

Alavijeh KP and Sharma D. 2012. *A study of antimicrobial activity of few medicinal herbs.* Asian J Pl Sci Res. 2(4) 496-502.

Sharma A, Chandradraker S, Patel and Ramteke P. 2009. *Antibacterial activity of medicinal plants against pathogens causing complicated urinary tract infections.* Ind J Pharm Sci.71 (2)136-139.

Voravuthikunchai S and Supawita thanomjit, 2004. *Effective medicinal plants against entrohaemorrhagic Escherichia coli.* J Ethanopharm *94(49-54).*

Ushimaru PI, Barbosa LN, Fernandez A and Fernandez A. Jr. 2012. *In vitro antibacterial activity of medicinal plant extracts against Escherichia coli strains from human clinical specimens and interactions with antimicrobial drugs.* Nat Prod Res .26(16)1553-1557.

## Chapter 16

# Studies on phytoremediation using *Vetiveria zizanoides and Eichhornia crassipes* on sewage water

**C. R. Hima, M. Manimegalai and K. Shanthi**

PG and Research Department of Zoology
Kongunadu Arts and Science College
Coimbatore, Tamilnadu, India

### ABSTRACT

The plants *Vetiveria zizanoides* and *Eichhornia crassipes* are used for phytoremediation process of sewage water. Seawage water contains wastes from factories, human settlement wastes and also the wastes from nearby shopes. Phytoremediation process is conducted using these two plants against nutrients like phosphate, nitrate, magnesium, chloride, calcium, iron and silicate for a period of 10, 20 and 30 days. After treatment the nutrient level showed a significant decrease in sewage water. As a phytoremediation process the Vetiver and *Eichhornia* can reduce the leval of toxic nutrients because the morphological and physiological characters of both plants, more effective for controlling the nutrient pollution. So suggestion can be made that these two plants must be utilized as an efficient economical and ecological alternative to accelerate the removal and degradation of waste water pollutants.

## INTRODUCTION

A plentiful supply of clean water is essential for the survival of human beings, plants and animals. A water resource is one of the major components of environmental resources that are under threat either from over exploitation of pollution, exhibited by human activities on the earth surface (Efe, 2005). There are several reports on the pollution of different fresh water bodies and studies on the chemical and physical parameters of several drinking water samples from surface water (Keshwani *et al*., 2004).

Due to man's exploitation, water resources have become darkened, smelly choked with excessive growth of algae. Aerobic bacteria decompose organic matter deplete $O_2$ content in water and fail to support fauna and flora (Indra Raj, 2000). Both urban and industrial complexes grow rapidly in the neighbourhood of rivers, lakes and ponds as water is the basic raw material for their survival, in ancient times conservation of water resources were considered as a basic law of self preservation (Elayarajan and Boopathi, 2006).

Industrial effluents constitute a major source of water pollution, in conjunction with other sources. According to the world commission on water for the 21$^{st}$ centuary, more than half of the world's major rivers are so depleted and polluted that they endanger human health and posion the surrounding ecosystems.

Phytoremediation is a promising clean up technology for waste water, ground water that is both low technology and low cost. The most important factor in implementing phytoremediation is the selection of an appropriate plant. Vetiver grass is effective in all its applications because it has a very large and deep penetrating root system that can grow to about 4 meters. The tensible strengh of vetiver grass is about one eighth of mild steel, which allows it to be so resistant to environmental influences (Cheng *et al.,* 2003). Water hyacinth has been used in aquatic systems for waste water purification in many years world wide. It is considered as the most efficient aquatic plant used in removing vast range of pollutants such as organic matters nutrients and heavy metals (Jayaweera *et al.*, 2004).

With the above views in mind an attempt has been made to analyase the leval of nutrients by phytoremediation using vetiver and water hyacinth in sewage water.

## MATERIALS AND METHOD

In the present study sewage water was collected from the drainage in front of the college campus. Drainage water contains the waste water coming from near by shops, factory and human settlement wastes. Sample was collected in a clean plastic container. This sample was brought to the laboratory and was analysed for nutrient levals. (Calcium, Magnesium, Nitrate, Phosphate, Iron, Silicate, Chloride).

### The following methods were used to estimate the minerals such as chloride, phosphate, nitrate, silicate, iron, calcium and magnesium:

Chlorides: Mohr's method (Jackson, 1973).
Phosphate: Strickland and Persons (1965)
Nitrate : Jingaran *et al.* (1969).
Silicate : Jingaran *et al.* (1969).
Iron : Stricland and Parsons (1965)
Calcium : Jingran *et al. (*1969)
Magnesium : Jingran *et al*. (1969).

### Collection of Vetiver and Water hyacinth Vetiver (*Vetiveria zizanoides*)

Healthy plants of Vetiver with roots were procured from Tamil Nadu Agricultural University, Coimbatore. The collected plants were washed thoroughly with tap water and then with distilled water to remove the soil particles and other contaminants. This plant is used for phytoremediation process.

## Water hyacinth (Eichhornia crassipes)

Water hyacinth was collected from Ukkadam pond, Coimbatore District. The collected plants were washed to remove the contaminants and kept in fresh water. These *Eichhornia* plants were used for phytoremediation.

## Experimental setup

The nutrients present in the sewage water sample were measured and taken as control. 2.5 litre of the water was taken in 2.5 litre capacity plastic containers. Vetiver and water hyacinth. The experimental set up was kept for 30 days in an undistrubed place and the nutrient content were analysed after 10, 20 and 30 days of exposure. Values were recorded and the percentage of increase or decrease over control was calculated.

**Table 1. Effect of Vetiver and Water hyacinth on mineral level in Sewage water after 10 days**

| Nutrients | Control Mg/ml | Vetiver | % of decrease | Water hyacinth | % of decrease |
|---|---|---|---|---|---|
| Phosphate | 4.9 | 3.6 | 26 | 4.1 | 16 |
| Nitrate | 574 | 432 | 24 | 485 | 15 |
| Magnesium | 380 | 265 | 30 | 285 | 25 |
| Chloride | 41.89 | 29.65 | 28 | 34.42 | 17 |
| Calcium | 225 | 162 | 28 | 185 | 17 |
| Iron | 2.40 | 1. 79 | 25 | 2.00 | 16 |
| Silicate | 5.17 | 4.01 | 29 | 4.85 | 15 |

**Table 2. Effect of Vetiver and Water hyacinth on mineral level in Sewage water after 20 days**

| Nutrients | Control Mg/ml | Vetiver | % of decrease | Water hyacinth | % of decrease |
|---|---|---|---|---|---|
| Phosphate | 4.9 | 2.8 | 42 | 3.1 | 36 |
| Nitrate | 574 | 312 | 45 | 352 | 38 |
| Magnesium | 380 | 196 | 48 | 225 | 40 |
| Chloride | 41.89 | 20.99 | 49 | 24.85 | 41 |
| Calcium | 225 | 112 | 50 | 138 | 38 |
| Iron | 2.40 | 1.21 | 49 | 1.42 | 40 |
| Silicate | 5.17 | 3.25 | 43 | 3.62 | 36 |

**Table 3. Effect of Vetiver and Water hyacinth on mineral level in Sewage water after 30 days**

| Nutrients | Control Mg/ml | Vetiver | % of decrease | Water hyacinth | % of decrease |
|---|---|---|---|---|---|
| Phosphate | 4.9 | 1.3 | 73 | 2.1 | 57 |
| Nitrate | 574 | 192 | 66 | 240 | 58 |
| Magnesium | 380 | 102 | 73 | 186 | 51 |

| Chloride | 41.89 | 12.24 | 70 | 19.15 | 54 |
|---|---|---|---|---|---|
| Calcium | 225 | 56 | 75 | 96 | 57 |
| Iron | 2.40 | 0.78 | 67 | 0.95 | 60 |
| Silicate | 5.71 | 1.85 | 67 | 2.18 | 61 |

## RESULTS AND DISCUSSION

Phosphate is an important mineral and it is reported to be the limiting factor for algal growth. In sewage water, (control) phosphate was noted as 4.9 mg/l. After Vetiver treatment the percentage of decrease over control was found to be 26, 42 and 73% and after *Eichhornia* treatment it is 16, 36 and 57% in 10, 20 and 30 days.

The main source of nitrate is the decomposition and biodegradation of organic matters. In sewage water, (control) nitrate value was noted as 574 mg/l. After Vetiver treatment, the percentage of decrease over control was found to be 24, 45 and 66% and after *Eichhornia* treatment it is 15, 38 and 50% in 10, 20 and 30 days.

Magnesium is similar in respect to calcium and usually present in solution as bicarbonate. It is absolutely essential for chlorophyll bearing organisms. In sewage water (control) magnesium value was noted as 380 mg/l. After Vetiver treatment the percentage of decrease over control was found to be 30, 48 and 73% and after *Eichhornia* treatment it is 25, 40 and 51% in 10, 20 and 30 days.

Chloride has been a basic parameter of detecting pollution of water by sewage before the development of bacteriological procedure. In sewage water, control chioride value was noted as 41.89 mg/l. After Vetiver treatment the percentage of decrease over control was found to be 28, 49 and 70% and after *Eichhornia* treatment it is 17, 41 and 54% in 10, 20 and 30 days.

Calcium is abundant in fresh water and it contributes hardness to water. In sewage water, control calcium value was noted as 225 mg/l. After Vetiver treatment the percentage of decrease over control was found to be 28, 50 and 75% and after *Eichhornia* treatment it is 17, 38 and 57% in 10, 20 and 30 days.

In the sample the control iron value was noted as 2.40 mg/l. After Vetiver treatment the percentage of decrease over control was found to be 25, 49 and 67% and after *Eichhornia* treatment it is 16, 40 and 60% in 10, 20 and 30 days.

In sewage water the control silicate value was noted as 5.71 mg/l. After Vetiver treatment the percentage of decrease over control was found to be 29, 43 and 67% and after *Eichhornia* treatment it is 15, 36 and 61% in 10, 20 and 30 days.

From the above syudy it is clear that these two plants are efficient for absorbing the nutrients. This may be due to the high absorbtion capacity of vetiver root against nutrients. The similar observations are made by Shu (2003) and Troung *et al.* (2000) *Eichhornia* is also a good nutrient absorber and it can help to reduce eutrophication and improve the water quality.

## REFERENCES

Chen, F.Y., Li, F. and Cheng, H., (2003). *A study on introduction test of vetiver grass*. Bull. Soil and Water Conservation, 11(3): 60-64

Efe, S.I., Ogban, F.E., Horsfall, M. Akporhonor, E.E., (2005). *Seasonal variation of physico-chemical characteristics in water resources quality system Western Nigeria delta region, Nigeria.* J. Appl. Sci. Environ. Mgt .9(1) : 191-195.

Elayarajan, M. and Murugesa Boopathi, P., (2006). *Effect of paper factory effluent on germination and seedling vigour of selected crops.* Indian J. Ecobiol. 19(2): 173-183.

Indra Raj, (2000). *Issues and objectives in ground water quality monitoring programme center hydrology project.* Proc. Natl. Symp. Ground water quality monitoring, Banglore. 1-7.

Jayaweera, M.W and Kasturiarachuchi J.C, (2004). *Removel of nitrogen and phosphorus from industrial waste waters by phytoremediation using water hyacinth* . Journal of Waste Water Science and Technology. 50(6): 217-225.

Kesharawani, Sadhana, A.K., Mandoli and Dube, K.K., (2004). *Determination of water quality index (WQI) of Amkhera pond of Jabalpur city.* National J. Life Sci., 1(1):61-66

Shu, W.S., (2003). *Exploring the- potential utilisation of vetiver in treating and mine drainage (AMD).* Proceedings of ICV- 3.

Troung, P.N. (2000). *The global impact of vetiver grass technology on the environment.* Proc. Second international vetiver conference, Thailand.

Chapter 17

# *In-vivo* analgesic activity of methanolic leaf extract of *Molineria recurvata* on albino rat

**Prasanta Dey[1], Manidipa Mukherjee[2], Ranjib Ghosh[3], Rajkumar Majumder[3], Tejendra Bhakta[2*]**

[1]C.L.Baid Metha College of Pharmacy, Thoraipakkam, Chennai, India

[2]Regional Institute of Pharmaceutical Science & Technology, Agartala, Tripura (W), India

[3]Tripura Medical College and BRAM teaching Hospital, Hapania, Tripura (W), India

**ABSTRACT**

The methanolic extract of *Molineria recurvata* was screened for its analgesic activity using the tail flick method in a dose of 200 mg/kg on albino rat which is compared with the standard drug (nimesulide). The result proves that the extract exhibit sufficient analgesic activity by central mechanism.

## INTRODUCTION

Pain is an unpleasant sensation which is associated with actual and potential tissue damage which may also be sensory and emotional (Satoskar *et al.*, 2001). Analgesics are those drugs which selectively relieves such pain by acting in the CNS or on peripheral pain mechanisms, without significantly altering consciousness (Tripathi, 2010).

*Molineria recurvata* is a herb (also known as palm grass). Leaves are alternate having parallel-veined, or pinnately veined. Stomata are present. Secretary cavities may be present or absent. Flowers are regular. This herb is available in the hilly areas of Tripura. The plant is well known for it's different folk medicines like leaves are used in bone fracture, in wound healing and as analgesic, anthelmintic (Prasanta Dey *et al,* 2012). The economic importance of this plant is, fibers that are collected from this plant have been used for different purposes such as making nets, and the fruit is edible.

The qualitative investigation test performed in five extracts (Pet. ether, Chloroform, acetone, benzene and methanol) revealed the presence of alkaloids, hydroxy-anthraquinone glycosides, steroids, carbohydrate and flavonoids etc. The presences of organic constituents are considered to be responsible for its different therapeutic actions. The aim of the work is to evaluate the analgesic activity of the said plant.

## MATERIALS AND METHODS

### Animals

Albino rat of either Sex (20 -25 g) were used for analgesic activity study (Chandrasekaran and Suseela, 2011). The animals were procured and housed in the animal house maintained under standard hygienic conditions at 20±2°C, humidity (60±10%). The experiment was done by using tail flick method[6]. The animals were divided into four groups each consisting of four. The groups were designated as follows.

Group 1 – control (saline water)
Group 2 –Nimesulide (5 mg/kg) i.p
Group 3 - 200 mg of methanolic extract of *M.recurvata*
Group 4 - 250 mg of methanolic extract of *M.recurvata*

### Oral Toxicology Test

The acute oral toxicity studies were performed to check the acute toxic effects and to determine minimum lethal dose of the drug extracts. The aqueous methanol extracts were administered orally to different groups of overnight fasted rat at the doses of 1000, 2000 and 3000 mg/kg body weight. After administration of the extracts, animals were observed continuously for the first three hours for any toxic manifestation. Thereafter, observations were made at regular intervals for 24 hrs.

### Tail flick method (Chandrasekaran and Suseela, 2011)

The tail of the mouse was placed on the hot wire (55$^{o}$C) and time taken by the animal to withdraw its tail from that hot wire was taken as the reaction time. Analgesic activity was measured at 30, 60 and 90 minutes after administration of the drugs. The analgesic activity was said to be complete if the mouse failed to withdraw its tail within 10 seconds of exposure.

**Table 1: Data showing the latency time for tail flicking at different time intervals (min) for various dosesof methanolic extract of *M.recurvata***

| Groups and treatment | Dose (mg/kg) | Response time after drug administration (Mean ± SE) in sec | | | |
|---|---|---|---|---|---|
| | | 0 mts | 30 mts | 60 mts | 90 mts |
| Control (saline) | - | 2.07 ± 0.035 | 2.18 ± 0.008 | 2.25 ± 0.005 | 2.32 ± 0.006 |
| Standard | 5 | 5.3 ± 0.057 | 7.23 ± 0.088 | 8.4 ± 0.057 | 5.63 ± 0.088 |
| Methanolic extract of *M.recurvata* | 200 | 5.06 ± 0.033 | 8.5 ± 0.057 | 9.42 ± 0.072 | 5.26 ± 0.073 |
| Methanolic extract of *M.recurvata* | 250 | 6.13 ± 0.088 | 8.83 ± 0.035 | 9.84 ± 0.032 | 6.30 ± 0.191 |

## RESULTS AND DISCUSSION

Methanolic extract of *M.recurvata* shows significant increase in basal reaction time which indicates positive analgesic activity shows in the table 1. The acute oral toxicity studies were performed to study the acute toxic effects and to determine $LD_{50}$ of the drug extracts. Albino rat were used for the study. The methanolic extracts were administered orally at the doses of 200 and 250 mg/kg body weight and observed for 24 hours. But no rat behavioral change is observed neither anyone was killed by the dose. Hence it is proved that the drug has no toxic effect in this level.

The qualitative investigation test proves the presence of alkaloids, hydroxy-anthraquinone glycosides, steroids, carbohydrate and flavonoids etc. Among these alkaloid may be responsible for its analgesic activity. Further study is under progress to isolate the pure component fraction. From the above experiments, it can be safely concluded that the polyherbal extract of *M. recurvata* indeed have a distinct analgesic property and it may act by elevating the pain threshold in the body.

## ACKNOWLEDGEMENTS

Authors are grateful to department of Pharmacology, Tripura Medical College and BRAM teaching Hospital, Hapania, Tripura (W), India for providing us the laboratory facilities.

## REFERANCES

Satoskar, R.S, S.D.Bhandarkar, S.S.Ainapure, (2001). *Pharmacology and Pharmacotherapeutics* 17th edition, Popular Prakashan Private Limited, Mumbai, p-135.

Tripathi, K. D, (2010). *Essentials of medical pharmacology*, 6th edition, Jaypee Brothers Medical Publishers (P) Ltd, New Delhi, p-453.

Prasanta Dey, Prasenjit Debnath, Tejendra Bhakta, (2012). *Evaluation Of Anthelmintic Activity Of Molineria Recurvata Leaf Extracts.* Int. Res J Pharm. App Sci.: 2(2):16-20.

Chandrasekaran, K. and Suseela, L. (2011). Analgesic *Activity Of Ethanolic Extract Of Pentatropis Capensis And Sarcostemma Secamone.* Journal of Pharma and Bio Sciences, Vol 2(4) :73-76.

## Chapter 18

# Conservation and management aspects of the sacred groves in Mahe, U.T. of Puducherry

**K. Sasikala, G. Pradeepkumar, C. C. Harilal[1] and C. P. Ravindran**
P. G. Department of Plant Science,
Mahatma Gandhi Govt. Arts College, Mahe, U.T. of Puducherry
[1] University of Calicut, Kozhikode, Kerala

**ABSTRACT**

The plant diversity, socio-cultural aspects, conservation and management of 19 sacred groves in Mahe were studied and documented. A total of 324 species of flowering plants and 18 species of non-flowering plants were recorded. Attempts were also made to document the conservation and management of the sacred groves. This paper discusses in detail the threats faced by the groves and strategies for conservation and sustainable management.

## INTRODUCTION

Sacred groves are patches of relict vegetation that are usually associated with folk deities and are conserved by the local communities based on taboos, religious beliefs and social sanctions (Gadgil & Vartak, 1975, 1976). About 17,000 sacred groves have been reported all over India and are associated with temples, monasteries/shrines or burial grounds and are found all over the country. They are referred to by different names in the country - Deorali in Darjeeling Hills, Lakyntang in Khasia and Jaintia Hills in Meghalaya, Jankor in Bihar and Madhya Pradesh, Oran in Rajasthan, Deorais or Deoranis in Maharashtra, Devarakadu, Devarabana or Sidharavanam in Karnataka, Kovilkadu or Kavu in Tamil Nadu and Sarpakavu or Kavu in Kerala.

The sacred groves of north Kerala are subjected to detailed study by various botanists. Jayarajan (2004) documented the biodiversity of the sacred groves of North Malabar. Sacred Groves of North Kerala was studied by Unnikrishnan (1995). Pushpangadan *et al.* 1998 studied the sacred groves of Kerala in general. Biodiversity of 123 sacred groves found in various parts of Puducherry, has been documented. King (1997), Kadamban (1998), Ramakrishnan *et al.* (1998), Praveenkumar (1999), Ramanujam & Kadamban (1999), Ramanujam *et al.* (2002), Ramanujam & Cyril (2003), Krishnan (2004), Parthasarathy *et al.* (2005) have documented the flora of the sacred groves from the Pondicherry region. Praveenkumar & Ramanujam (2003) documented the belief system and people's perceptions of sacred groves in Puducherry. Devaraj *et al.* (2005) reported the status of sacred groves of Pondicherry region and discussed the need for conservation. Recently, Jisha (2005) documented the phytodiversity of the five groves of Mahe region. Ramanujam *et al.* (2007) made a comprehensive floristic survey of the sacred groves of Puducherry region. A preliminary survey of the flora of Mahe has been studied by Sasikala *et al.* (2009 a, b & c). In all these studies a comprehensive account pertaining to plant diversity, socio-cultural aspects, conservation and management is lacking. Hence the present study was carried out.

Mahe under the Union Territory of Puducherry occupies a unique geographical location near Arabian Sea. It is situated on the West Coast of the Indian Peninsula between 11°42' - 11°43' N and 75°31' - 75°33' E. It covers an area of 9 sq.km. Mahe also forms a part of the Western Ghats, one of the hottest hotspots in the world.

## MATERIALS AND METHODS

Intensive field surveys were conducted during the year 2010 - 2012 to study the sacred groves. A questionnaire has been prepared to facilitate the collection of data pertaining to the historical, cultural and existing conservation practices of each sacred grove. The taboos, rituals and beliefs associated with the groves were recorded with the help of priests, elderly people and the stake holders of the groves. Observations on the diversity and distribution of endemic and threatened plant species in sacred groves were made. Plant specimens collected during the study are identified using relevant floras (Ramachandran & Nair, 1988; Sasidharan, 2004; Nayar *et al.*, 2006) revisions, monographs and research articles. Plant materials made into herbarium specimens following standard herbarium techniques proposed by Fosberg & Sachet (1965) and are deposited at herbarium of Mahatma Gandhi Government Arts College, Mahe, for reference.

## RESULTS AND DISCUSSION

A total of 19 sacred groves (Kavu) have been located in Mahe. These include 5 each situated in Pandakkal and Palloor and 3 each in Chembra, Chalakara and Mahe Proper. Pando Kavu Sri Ayyappan Temple, Panolil Kavu and Chembra Sri Ayyappan Kavu are considered major groves and the remaining 16 are recognized as minor groves. The sacred groves predominantly harbour various life-forms of angiosperms such as herbs, shrubs, trees, climbers, epiphytes and parasites. Preliminary floristic surveys conducted in 19 different sacred groves of Mahe recorded a total of 324 angiosperm species belonging to 95 families. These include 257 species of dicots and 67 monocots. The life-form composition analysis shows that herbs dominate the flora of sacred groves with a total of 138 species (42%), followed by trees with 100 species (31%), shrubs with 55 species (17%) and climbers with 31 species representing 10% of the flora. A total of 14 endemic species have been recorded from the sacred groves. Among the 19 sacred groves that have been floristically surveyed in Mahe, the maximum species diversity was observed in Pandokavu and the least in Sarpa Kavu. The species composition indicates the remnants of tropical evergreen forest which might have existed in the past in this region.

The sacred groves are the sites for socio-cultural and spiritual interactions. The owners of the kavu take the responsibility of conducting the rituals. The traditional customs, rites, rituals, taboos, folklore and faith creates an emotional bond between culture and conservation. They respect these taboos with much sincerity even today. They observe rituals and organize festivals to please the deities and to ward off dangers in the form of sickness, epidemics and demanding good health. If they violate the laws of sacred groves they will have to face untoward incident due to wrath of deity. Daily pooja is performed only in Pandakkal Ayyappan Kavu and Chembra Ayyappan Kavu whereas in other groves no daily pooja is performed, instead lighting of the lamps is done in the morning and evening. Offerings such as money, flowers, coconut oil are made at all times.

Theyyam festival has been celebrated once in every year. Theyyam is the unique ritual related to kavus of northern Kerala in Malabar region and is a part of the annual festivals of all kavus to renew the deities. Generally in Ayyappan and Sarpakavu there is no Theyyam. Theyyam is performed with the help of Vadyagosham except in Sri Nelliyat Kalari Bhagavathy Kavu where neither Theyyam nor Vadhyagosham is performed. There is no animal sacrifice except in Puthalam Temple and Sri Peradayath kavu where hens are sacrificed to propitiate/appease the deity. Few sacred groves are associated with pond or rivulets which act as micro-watersheds. The performers are decorated with costumes and magnificent make up and gorgeous colours. A strict code of conduct has been observed by the owners of kavu and local people in protecting the sanctity of sacred groves.

Sacred groves have been conserved as sustainable biological resources. They serve as a valuable gene pool. They are also considered as the first major effort to recognize and conserve biodiversity. Since ancient period kavu has been recognized as the source of medicine, edibles and other economically important plants for the natives. The notable medicinal species include *Anamirta cocculus*, *Curculigo orchioides*, *Cyclea peltata*, *Gloriosa superba*, *Glycosmis pentaphylla*, *Hydnocarpus pentandrus*, *Ocimum* spp., *Phyllanthus amarus*, *Sarcostigma kleinii*, *Scoparia dulcis* and *Tinospora cordifolia*.

## CONCLUSION

Sacred groves are an abode of good number of rare and threatened native species of flora and fauna. Preservation of sacred groves is one of the best methods for maintaining the biodiversity. Beliefs system and taboos associated with the groves help in conservation of biodiversity. However, many sacred groves are now threatened and altered both in form and size; similarly, the vegetation and species composition of sacred groves have also been altered due to change in socio-economic condition and land use system. The impact of threats varies from region to region as well as between groves, accordingly the level of degradation in each grove.

The documentation of the sacred groves will help to assess the status of the groves, their biodiversity, environment, traditional culture and management. The present study establishes the biological richness of the sacred groves of Mahe. Of the 19 sacred groves, Pando Kavu, Chembra Ayyappan Kavu and Panolil Kavu are larger in size compared to others and are also well protected and conserved. Some of the groves have been maintained in good condition in spite of being situated closer to the human habitations. It has been strongly believed by grove authorities that the untoward incidents happened in their families are considered the aftermaths of curse by the deities of the sacred groves as they were not properly maintained by the owners. Therefore, the authorities are taking necessary effort to revive the quality of the sacred groves. The restoration of such groves will be done based on ecological knowledge gained from well preserved groves like Sri Chembra Ayyappan Kavu and Panolil Kavu.

Major threats to the groves and their biodiversity include habitat destruction or degradation due to a number of developmental activities, acquisition of lands for construction of buildings and roads and invasion of alien species. Necessary efforts need to be taken to protect the sacred groves to prevent the loss of biodiversity. Clearing of ground flora during the festival time drastically affect the diversity of herbs and regeneration of arboreal species. Plantation of native species in the groves, especially tree species need to be encouraged. Introduction of alien invasive species in the groves should be banned.

By conducting appropriate management practices and environmental awareness programmes among the stake holders and local populace, and also by giving incentives to the people who are involved in the protection of the grove, many of the groves can be maintained and preserved effectively in their existing condition without any loss of biodiversity. However, every individual should take the responsibility of saving the sacred groves for their traditional, cultural, aesthetic and ecological values. The data generated in the present study can provide a lot of inputs to the authorities who prepare management plan, the researchers and academicians for the conservation of the sacred groves of Mahe.

## ACKNOWLEDGEMENTS

Authors are thankful to the Department of Science, Technology and Environment, Govt. of Puducherry, Puducherry, for financial assistance. They are highly grateful to the Principal, for encouragement and Dr. W. Arisdason, Scientist, Botanical Survey of India, Kolkata, for valuable suggestions and carefully going through the manuscript.

## REFERENCES

Devaraj, P., Ramanaujam, M.P. & Ganesan, T. (2005). *Status Report of Sacred Groves of Pondicherry Region and Strategies for Conservation.* In: Kunhikannan, C. & Singh, B.G. (eds.), Proc. Natl. Workshop Strat. Conserv. Sacred Groves. Institute of Forest Genetics and Tree Breeding, Coimbatore. pp. 16 - 21.

Fosberg, F.R. & Sachet, H. (1965). *Manual of Tropical Herbaria.* Regnum Veg. Vol. 39.

Gadgil, M. & Vartak, V.D. (1975). *Sacred groves of India - A plea for continued conservation.* J. Bombay Nat. Hist. Soc. 72(2): 314 - 320.

Gadgil, M. & Vartak, V.D. (1976). *The Sacred groves of Western Ghats in India.* Econ. Bot. 30: 152 - 160.

Jayarajan, M.( 2004). *Sacred Groves of North Malabar.* Kerala research Programme on Local level Development, Centre for Development Studies, Thiruvananthapuram.

Jisha, V.C. (2005). *Sacred Groves of Mahe.* M. Phil. Dissertation (unpublished), Pondicherry University, Puducherry.

Kadamban, D. (1998). *Biocultural Perspectives and Plant Diversity of Sacred Groves and Traditional Medical Knowledge in Pondicherry Environs.* Ph. D. Thesis (unpublished), Pondicherry University, Puducherry.

King, O.I. 1997. *Botanico and Socio cultural Studies on Sacred Groves in Cuddalore District, Tamil Nadu.* M. Phil. Dissertation (unpublished), University of Madras, Chennai.

Krishnan, V. (2004). *Plant Diversity and Biocultural perspectives of sacred groves in Pondicherry and its environs.* Ph. D. Thesis (unpublished), Pondicherry University, Puducherry.

Nayar, T. S., Rasiya Beegam, A., Mohanan, N. & Rajkumar, G. (2006). Flowering Plants of Kerala - A Handbook. Tropical Botanic Garden and Research Institute, Thiruvananthapuram.

Parthasarathy, N., Venkateswaran, R., Reddy, S. & Mani, S. (2005). *Role of Sacred Groves in Biodiversity conservation of Tropical Dry Evergreen Forests.* In: Kunhikannan, C. & Singh, B.G. (Eds.), Proc. Natl. Workshop Strat. Conserv. Sacred Groves. Institute of Forest Genetics and Tree Breeding, Coimbatore. pp. 40 - 46.

Praveenkumar, C. (1999). *Plant Biodiversity and Biocultural Traditions of Four Sacred Groves of composite South Arcot District of Tamil Nadu.* M. Phil. Dissertation (unpublished), Pondicherry University, Puducherry

Praveenkumar, C. & Ramanujam, M.P (2003). *Belief system and people's perceptions of our Sacred Groves in the Pondicherry of South India.* ENVIS - Newsletter, CPRI, Chennai

Pushpangadhan, P., Rajendraprasad, M. & Krishnan, P.N.(1998). *Sacred Groves of Kerala – A synthesis on the state of art of knowledge.* In: Ramakrishnan, P.S., Saxena, K.G. & Chandra Shekara, U.M. (Eds.), Conserving the Sacred Groves for Biodiversity Management. New Delhi. pp. 193 - 210.

Ramachandran, V.S. & Nair, V.J. (1988). *Flora of Cannanore.* Botanical Survey of India, Calcutta.

Ramakrishnan, P.S., Saxena, K.G. & Chrandrashekara, U.M. (1998). *Conserving the Sacred Groves for Biodiversity Management.* Oxford, New Delhi.

Ramanujam, M.P. & Cyril, K.P.K. (2003). *Woody species diversity of four sacred groves in the Pondicherryregion of South India.* Biodivers. & Conservation 12: 289 - 299.

Ramanujam, M.P. & Kadamban, D. (1999). *Rare, Endangered and Threatened plants occurring in the sacred groves of Pondicherry region.* Natl. Symp. Emerging Trends Pl. Sci. Sri Venkateswara University, Tirupathi. p. 43.

Ramanujam, M.P., Kadamban, D., Kumaravelu, G. & Praveenkumar, K. (2002). *Sacred Groves – An overview* .In: Trivedi, P.C. (ed.), Ethnobotany. Jaipur. pp. 13 - 53.

Ramanujam, M.P., Ganesan, T., Kadamban, D., Kumaravelu, G. & Devaraj, P. (2007). *Flora of Sacred Groves of Puducherry (A Pictorial Guide).* Dept. of Forest & Wildlife, Puducherry.

Sasidharan, N. (2004). *Biodiversity Documentation for Kerala.* Part 6: Flowering Plants. Kerala Forest Research Institute, Peechi.

Sasikala, K., Girishkumar, E. & Krishnaveni, C. (2009 a). *A preliminary survey of the wild edible plants of Mahe, U.T. of Puducherry..* Proc. National Seminar on Recent Trends in the conservation and utilization of underutilized wild edible plants. Bharathiar University, Coimbatore, 158 – 163.

Sasikala, K., Rahina, K. & Arisdason, W. (2009 b). *Exotic flora of Mahe, U.T. of Puducherry.* Eco-chronicle 4(2): 265 - 271.

Sasikala, K., PradeepKumar, G., Arisdason, W., Girish Kumar, E. and Ravindran, C.P. (2009 c). *A Preliminary Analysis of the Flora of Mahe, Union Territory of Puducherry.* Eco-chronicle 4(4): 219 - 228.

*Unnikrishnan*, E. (1995). *Sacred Groves of North Kerala – An Eco-folklore Study.* Thrissur. (in Malayalam).

Chapter 19

# Mosquito larvicidal activity *of Momordica charantia* leaf extract against filarial vector *Culex quinquesfasciatus*

**Annapoorani C. A. and Manimegalai K.**
Department of Zoology
Avinashilingam Institute for Home Science of Higher Education for Women
Coimbatore

**ABSTRACT**

Mosquitoes serve as obligate intermediate host and transmit many deadly diseases. It is also an important pest that infects human beings. Many mosquito species developed resistance against synthetic insecticides. Plants possess phytochemicals, which is an alternative to synthetic insecticides. Plants possess phytochemicals, which can be used as an alternative to synthetic insecticides and effective in mosquito control. Plant extracts avoid adverse toxic effects on environment and prevent pollution. In the present study, larvicidal activity of leaf extracts of *Momordica charantia* against the fourth instar larvae of *Culex quinquefasciatus.* Results showed that petroleum ether extract was found to be more effective than chloroform and ethanol extracts.

## INTRODUCTION

Mosquitos have greater influence on human health and well-being throughout the world than any other arthropod, mainly because of their involvement in both transmitting a number of dreadful diseases such as malaria, filarial, dengue, chikungunya etc, which create nuisance to public health. These medically important mosquitoes which act as a vector for the transmission of serious diseases that causes morbidity, mortality, economic loss and social disruption (Becker, 2003).

The control of these diseases is largely dependent on spraying of chemical insecticides to kill mosquito adults or larvae. Using the larvicidal activity is an effective method to reduce the mosquito densities before they emerge as adults and synthetic insecticides have been widely used for this purpose (Tiwary *et al.*, 2007). The use of these synthetic pesticides has been found to have side-effects in non-targeted organisms (Bernadou *et al.*, 2009). Natural products of plant origin with insecticidal properties have been tried in the recent past for control of variety of insect pest and vectors. Phytochemicals derived from plant sources can act as larvicides, insect growth regulators and repellent and ovipositor attractant and have different activities (Venketachalam and Jebasan, 2001).

## MATERIALS AND METHODS

### Laboratory culture of larvae

Hay infusion method was adopted for culturing mosquito larvae. Hay was taken, cut into small pieces and boiled in 5 litres of water for 20 minutes. After cooling, this water was poured into buckets and kept in different areas where mosquitoes were abundant. After one or two days eggs were laid by female mosquitoes in clusters forming an egg raft. The egg rafts were collected and maintained in the laboratory. The third instar larvae were collected, reared in enamel trays containing culture medium and provided with powdered dog biscuits and yeast in the ratio of 3:1 as the nutrient source. Immediately after moulting, the fourth instar larvae were introduced into beakers containing 200ml of water and used for the bioassay studies.

### Preparation of leaf and seed extracts

Fresh leaves of *Momordica charantia* were collected, washed in water and air dried under shade. Dried leaves and seeds were powdered using an electric pulverizer. Fine powder was obtained by sieving. 10g of the seed powder was weighed and made into packets using Zerohaze filter paper. This powder was subjected to extraction with 500ml of the solvents such as petroleum ether, chloroform, and ethanol for 8h using a soxhlet apparatus. Petroleum ether (60-80$^{\circ}$C) extraction was followed by chloroform extraction and ethanol extraction, so that the powders were subjected to extraction with solvents of increasing polarity. The leaf extract thus obtained was concentrated by distillation and dried by evaporation at 40$^{\circ}$C.

### Bioassay studies

A pilot study was carried out to find out the effective doses of leaf extracts that produced mortality in larvae of *Culex quinquefasciatus*. After determining the effective doses, the detailed investigations were undertaken. Concentrations ranging from 20-120 ppm were used which produced larval mortality at or after 24 h treatment. Hence, 20-120 ppm concentrations were used for the bioassay studies.

### Experimental design

The experiment was laid down in a Completely Randomized Design (CRD). The experimental set up consisted of six treatments each with three replications, one set for leaf extract. Twenty newly emerged fourth instar larvae of *Culex quinquefasciatus* were introduced into the breakers for the bio-assay studies. Controls with three replicates were also maintained simultaneously. 250 ml beakers each containing 200ml of water were used for the experiments. The following parameters were observed to assess the effective dose. Larval mortality at 24h, 48 h, 72 h and 96 h of treatment, larval and pupal behavior, pupal mortality, total developmental duration, morphological changes, incidences of malformations and adult emergence were observed.

### Test for larvicidal activity

The larval mortality in both treatment and control was recorded at 24h of treatment and the percentage of mortality was calculated using Abbott's formula (Abbott, 1925).

$$\% \ mortality = \frac{Mortality \ in \ treatment \ (\%) - Mortality \ in \ control \ (\%)}{100 - Mortality \ in \ control \ (\%)}$$

## RESULT AND DISCUSSION

Benzene, hexane, ethyl acetate, methanol and chloroform leaf extract of *Andrographis paniculata* werefound to be more effective against *C. quinquefasciatus* than *A. aegypti* (Govindarajan, 2011). Mohan *et al.*, (2005) observed that the fruits of *S. xanthocarpum* were very effective against the larvae of *A. stephensi* and *C. quinquefasciatus*.

Essential oil of *Ipomoea cairica* possesses remarkable larvicidal properties as it could produce 100% mortalilty in the larvae of *C. tritaeniorhynchus, C.quinquefasciatus, Ae.aegypti,and An. stephensi* mosquitoes at concentrations ranging from 100 to 170 ppm (Thomas et al., 2004). The larval mortality recorded at 24 hr was comparatively more in higher doses (400 ppm) of ethanol extract (31.66%) as compared to chloroform and petroleum ether extracts. (Table1) Maximum mortality of 46.66% was observed in petroleum extract at 48 hr while in chloroform extract the maximum larval mortality was 25% and in ethanol extract it was 6.66%

## CONCLUSION

From this study, the presence of phytochemicals in the leaf of *Momordica charantia* might be the reason for its larvicidal activity. The results of this experiment indicated that extracts of the *Momordica charantia* can be used as environmental friendly and sustainable insecticides to control mosquitoes.

**Table 1 showing the effect of various extracts of leaves of *Momordica charantia* on mortality of mosquito larvae**

| Extracts used | Hour | Extract concentration in ppm | | | | | | |
|---|---|---|---|---|---|---|---|---|
| | | Control | 330ppm | 320ppm | 340ppm | 360ppm | 380ppm | 400ppm |
| Petroleum ether | 24 | 0 | 13.33 | 11.66 | 15.00 | 15.00 | 20.00 | 26.66 |
| | 48 | 0 | 35.00 | 38.33 | 40.00 | 38.33 | 41.66 | 46.66 |
| | 72 | 0 | 10.00 | 6.66 | 11.66 | 21.66 | 36.66 | 26.66 |
| | 96 | 0 | 5.00 | 8.33 | 13.33 | 23.33 | 1.66 | 0 |
| | Pupal mortality | 0 | 6.66 | 6.66 | 0 | 0 | 0 | 0 |
| | Total mortality | 0 | 70.00 | 71.66 | 78.33 | 98.33 | 100 | 100 |
| Chloroform | 24 | 0 | 13.33 | 13.33 | 15 | 15 | 15 | 18.33 |
| | 48 | 0 | 13.33 | 13.33 | 15 | 20 | 23.33 | 25 |
| | 72 | 0 | 11.66 | 10 | 15 | 20 | 23.33 | 25 |
| | 96 | 0 | 11.66 | 11.66 | 13.33 | 21.66 | 21.66 | 16.66 |
| | Pupal mortality | 0 | 3.33 | 1.66 | 8.33 | 6.66 | 8.33 | 8.33 |
| | Total mortality | 0 | 53.33 | 51.66 | 66.66 | 76.66 | 81.66 | 86.66 |
| Ethanol | 24 | 0 | 5.00 | 10.00 | 15.00 | 11.66 | 18.33 | 31.66 |
| | 48 | 0 | 10.00 | 10.00 | 10.00 | 10.00 | 10.00 | 6.66 |
| | 72 | 0 | 15.00 | 15.00 | 10.00 | 13.33 | 11.66 | 8.33 |
| | 96 | 0 | 5.00 | 5.00 | 5.00 | 5.00 | 10.00 | 6.66 |
| | Pupal mortality | 0 | 1.66 | 10 | 0 | 0 | 0 | 0 |
| | Total mortality | 0 | 36.66 | 50 | 40 | 45 | 46.66 | 51.66 |

## REFERENCES

Abbott, W.S. (1925). *A method of computing the effectiveness of an insecticide*. Econ. Entomol: 18: 265-267.

Becer, N., Petriae, D., zgomba, M., Lane, J. and Kaiser,A.(2003). *Mosquitoes and their control.* New York: Kluwer Academic/Plenum Publishers.

Bernadou, A., Demares, F, Couretfauvel, T., Sandoz, J.C. and Gauthier, M. 2009. *Effect of fipronil on side specific antennal tactile learning in the honeybee.* J. Insect Physiol, 55. 1099-1106.

Govindrajan,M. (2011). Evaluation of *Andrographis paniculata* Burm.f. (Family: Acanthaceae) extracts against *C.quinquefasciatus* (Say) and *Ae. aegypti* (Linn) (Diptera: Culicidae). Asian pac J.Trop Biomed: 4(3): 176-181.

Mohan, L., Sharma,D. and Srivastava, C.N. (2005). *Evaluation of Solanum xanthocarpum extracts as mosquito larvicides* J.Environ, Biol., 26, 399-401

Thomas, T.g., Rao, S., Lal,S. (2004). *Mosquito larvicidal properties of an indigeneous plant, Ipomoea cairica Linn*: Japanese J insfect Dis: 57: 176-7

Tiwary, M., Naik, S.N., Tewary, D.K., Mittal, P.K.Yadav, S. (2007). *Chemical composition and larvicidal activities of the essential oil of Zanthoxylum DC (Rutaceae) against three mosquito vectors.* J Vec Borne Dis., 44: 198-204.

Venketachalam, M.R. and Jebasan, A. (2001). *Larvicidal activity of Hydrocotyl javanica Thunb (Apiaceae) extract against C. quinquefasciatus.* J. Exptl . Zool India 4(1) 99-101.

# Chapter 20

# Studies on larvicidal activity of *Gloriosa superba* on the development stages of the Chikungunia mosquito, Aedes aegypti

**C. R. Hima and M. Manimegalai**
PG and Research Department of Zoology
Kongunadu Arts and Science College
Coimbatore, Tamilnadu

**ABSTRACT**

In the present study the larvicidal effect of botanical insecticide, *Gloriosa superba* against *Aedes aegypti* was administrated. The plant extract showed higher mortality in lower doses and it exhibit better protection against the different developmental stages of *Aedes aegypti*. Based on the probit analysis, the LC50 values and 95% upper and lower fiducial limit and regression values of the plant extract on *Aedes aegypti* were also noticed. The treatment of seed extract against III and IV instar larvae of *Aedes aegypti* showed that III instar larvae are more susceptible than IV instar larvae. These morphogenetic abnormalities are commonly caused by botanical extract and the disturbance from the growth regulating hormones. It is because the active principle compounds such as tarpenoids, alkaloids, flavonoids, phenol and tannins present in this plant It is therefore suggested that *Gloriosa superba* derivatives can be effectively used as a biocide for vector control operations.

## INTRODUCTION

Mosquitos are the most important single group of insects well known for their public health importance as they transmit serious diseases to man. They not only cause nuisance by their bites but also act as vector for the pathogens of various diseases like malaria, Filariasis, Dengue fever, Chickunguniya and japenese encephalitis. They are most serious vector born diseases contributes significantly to poverty and social debility in tropical countries. *Aedes aegypti* is one of the most important diseases transmitting and nuisance causing mosquito transmits dengue and Chikunguniya, a serious mosquito borne epidemic as gained momentum in India.

The use of insecticidal sprays and other chemicals for killing the adult mosquitoes will affect the natural environmental conditions. Subthetic insecticides are toxic and adversely affect the environment by contaminating soil, air and water, so there is a constant need for developing biologically active plant materials as larvicides which is expected to reduce the hazards to human and other organisms by minimizsing the accumulation of harmful residues in the environment (Prabakar and Jebenasan, 2004).

In early historical times man used plant extract and compounds to control certain pests and disease causing insects. Plants are rich source of bioactive organic chemicals and synthesize a number of secondary metabolites that serve as insecticides, antifeedant, oviposition deferrents, adulticides, repellents, growth inhibitors. The natural products, especially plant derivative called bioinsecticides are one emerging as a viable component of integrated pest Management (IPM) strategies in view of their pesticidal potency as well as efficiency on pest and safety to parasitoids and predators (Kumbhar *et al.*, 2000).Hence the present study was carried out to find a biological way of solution using the plant extract of *Gloriosa superba* for the problem caused by the mosquito to the public.

## MATERIALS AND METHODS

For the present study the acetonic extract of seeds of *Gloriosa superba* were used to determine the groups of secondary metabolities present in it and to test the efficiency and insecticidal property against III and IV larval stages of the mosquito *Aedes aegypti*.

## EXPERIMENTAL DESIGN

### Test organisms

To satisfy the number of mosquitoes needed for the day to day bioassays, a colony was essential. The eggs of Aedes were procured from National Institute of communicable Disease (NICD) at Mettupalayam, India. The mosquito colony was maintained at 7.0 – 8.5 pH, 28+- $2^0$C temperature and 14:10 light and dark photo period cycle. The larvae were fed on powdered mixture of dog biscuits and yeast ratio 3:1, The pigeon blood meal was given to the female adult mosquitoes and 10% glucose solution was given to the male adult mosquitoes.

### Preparation of phytochemical extract

The seeds *Gloriosa superba* collected from field were brought to the laboratory and were dried under shade and were ground to fine powder. The100gm powder was extracted with 300 ml of acetone by using soxhlet apparatus for 8 hours (Vogel, 1978). The extract was concentrated in a vacuum evaporator at $45^0$C under low pressure. After complete evaporation of the solvent the concentrated extract was collected and stored in separate glass vessels at $4^0$C in refrigerator for further experiments. 25gm og the concentrated extract of dried seed of experiment plant were dissolved in 750 ml of acetone and kept as stock solution and were used to prepare the desired concentration for exposure of the mosquito larva.

### Larvicidal bioassay

The larvicidal bioassay was done using standard WHO protocols (WHO, 2005). To obtain the different concentrations of test medium, the stock solutions were dissolved in water and then mixed throughly with the dry ingredients of the diet. Twenty freshly moulted third and fourth instars larvae of Aedes aegypti were exposed to different concentrations of plant extracts. At each concentration 2 to 5 trials were maintained and each trial consist of four replications. Controls were maintained using 2 to 5 trials and each trial consist of four replications of different developmental stages. Observations were made over a period of 24 hours. The dead larvae were identified when then failed to move after probing with a needle in the siphon of cervical region. Mori bund larvae were those incapable of rising to the surface or showing the characteristic diving reactions when the water was disturbed.

## RESULTS AND DISCUSSION

There is a number of naturally occurring compound that posses plant protection properties. The botanical extracts from the plant, leaves, roots, seeds, flowers and bark in their crude form have been used as conventional insecticides for centuries. The preliminary screening is a good mean of evaluation of the potential larvicidal effects and the highest mortality was observed in the acetone seed extract of *Gloriosa superba* under laboratory conditions and is presented in the table 1

In the present study effect of acetone seed extract of *Gloriosa superba* were experimented against III and IV instar larvae of Aedes aegypti and the percentage of mortality was calculated. The percentage of mortality values for III and IV instar larvae of *Aedes aegyptio* treated with various concentrations of the seed extract of *Gloriosa superba* and LC50 values were presented table 2. Highest mortality of 87% and 72% was recorded for third and fourth instar larvae of Aedes aegypti at 40ppm concentration. Aedes aegypti larvae exhibited very little susceptibility to lower doses and the mortality of 27% and 19% was reported at the concentration of 30ppm.

The highest sensitivity of third instar of Aedes aegypti to the seed extract of *Gloriosa superba* was also evident by the lowest Lc50 values (LC50=34.62ppm). Least susceptibility was shown by fourth instar larvae of Aaedes aegypti (LC50=40.47ppm) respectively. Similar trend of differential susceptibility of mosquito species to plant extract was reported earlier (Pathak *et al.*, 2000). From the overall results it is clear that the seed extract of *G. superba* can be effectively used for the control of mosquito *Aedes aegypti.*

**Table 1: LC50 values of acetone seed extract of *Gloriosa superba* on III and IV instar of *Aedes aegypti***

| Larval Stages | % of Mortality | | | | | 95% Fiducial Limit (ppm) | | LC50 (ppm) | Regression Equation |
|---|---|---|---|---|---|---|---|---|---|
| | 30 ppm | 35 ppm | 40 ppm | 45 ppm | 50 ppm | LM | UM | | |
| III Instar | 27 | 39 | 51 | 60 | 72 | 34.62 | 43.15 | 34.62 | Y=-39+ 2.22x |
| IV Instar | 19 | 36 | 50 | 71 | 87 | 36.19 | 44.62 | 40.47 | Y=-84.2+ 3.42x |

## REFERENCES

Kumbhar, P.P., Salunkhe, D. H., Borse, M. B., Hiwale, M.S., Nikam, L.B., Mrs. R.S. Bendre., Kulkarni, M.V and Dewang, P.M. (2000). *Pesticidal potency of some common plant extracts.* Pestology. 24(6): 51-53.

Pathak, N, Mittal PK singh O.P Vidhya sagar, Vasudevan P (2000). *Larvicidal action of essiontal oils from plants against the vector mosquitoes Anopheles stephensi(liston), culex quinquefasciatus(say) and Aedes aegypti(L) Int.* Pest control. 42:53-58

Prabakar,K., Jabansan,A. (2004). *Mosquitocidal effect of bitter gourd lead extracts. Development of Zoology, Annamalai University.* Bio Resour, Technol.95(1) 113-114.

Vogel Al. (1978). *Text book of practical organic Chemistry*, the English language Book society and longman, London, P. 1368.

## Chapter 21

# Effect of green manure and biofertilizer on the biochemcial parameters of cluster beans (*Cyamopsis tetragonoloba* L. var. PNB)

**M. V. Anitha, B. Vijayakumari and S. R. Radha**

Department of Botany

Avinashilingam Institute for Home Science and Higher Education for Women

Coimbatore

**ABSTRACT**

In India agriculture is the main occupation of majority of the population. This versatile and primordial greenery around us plays a vital role in over all development of our country. An experiment was carried out in cluster beans (*Cyamopsis tetragonoloba* L. Var. PNB) at Avinashilingam Institute for Home Science and Higher Education for women, Coimbatore. Various combinations of organic treatments were used in the present study to assess the influence of herbals and biofertilizer on the biochemical parameters of *Cyamopsis tetragonoloba* L. Var. PNB. In each pot different green manure (Nochi, Turmeric, Vasambu, Goose berry, Kuppaimeni) and biofertilizer (Rhizobium) was mixed individually. The biochemical parameters of Cluster bean were recorded on 30, 45, 60 and 75 DAS. Chlorophyll 'a', 'b' and 'total' chlorophyll were higher in biofertilizer treatment ($T_6$) on all the days. The maximum protein content was shown on 30$^{th}$ day by $T_3$ as (Vasambu rhizome powder) 31.47 mg/g and on 45 and 60 DAS it was higher in $T_5$ (Gooseberry) (33.80 mg/g and 35.86 mgg). On 75 DAS it was higher in $T_6$ (biofertilizer) (38.52 mg/g). The highest carbohydrates was found in $T_6$ (biofertilizer) on 30 and 45 DAS (79.24 mg/g and 79.95 mg/g). On 60$^{th}$ day the highest carbohydrates content had been observed in $T_5$ (Gooseberry) 86.36 mg/g, on 75$^{th}$ day it was higher in $T_6$ (94.93 mg/g). The highest ascorbic acid content had been observed in $T_6$ (biofertilizer) on 30 DAS (40.00 mg/g), 45 DAS (80.00 mg/g), 60 DAS (246.66 mg/g) and 75 DAS (266.66 mg/g).

## INTRODUCTION

Agriculture is the core sector of Indian economy and is the backbone of more than 70 per cent of the Indian population. Agriculture has the power to change the economic and social condition of a nation. The agricultural development depends on a number of factors like natural, economical and social and sub factors like climate, soil, use of advanced technology, labour factor and fertility status. To have more production per unit of land we have to consider the fertility of the soil and amount of nutritive elements in the soil. If the nutritive elements are more or sufficient in soil, it is fertile, otherwise infertile. To make soil fertile, a nutritive ingredient is essential. Green manure is called a cover crop which adds nutrients to the soil. Green manure means planting a crop that is meant to be incorporated in to the soil to increase its fertility. Plant have cordial relationship with Rhizobium which form root nodules in legumes helps to improve nitrogen content of the soil. Hence in the present study the effect of various plant species were used as green manure and Rhizobium on *Cyamopsis tetragonoloba* L. Var. PNB were compared.

## MATERIALS AND METHODS

An experiment was conducted to assess the manurial value of the following botanicals (Nochi, Turmeric, Vasambu, Goose berry, Kuppaimeni) and biofertilizer (Rhizobium). The impact of soil treatment on the growth attributes, quality parameters and yield components of cluster bean (*Cyamopsis tetragonoloba* L. Var. PNB) was studied.

### Collection of various materials

### Collection of Soil Samples

Red sandy loam soil was collected from Saravanampaty, Coimbatore.

### Collection of Seeds

Seeds of cluster beans (*Cyamopsis tetragonoloba* L. Var. PNB) were procured from Department of Pulses, Tamil Nadu Agricultural University, Coimbatore.

The experiments were conducted at the Laboratory of Botany, Avinashilingam Deemed University, Coimbatore. A brief description of the test plant and plants used as manure is furnished below.

### Test crop

### Cluster Bean (*Cyamopsis tetragonoloba*) Seeds

Cluster bean is one of the most important legume crops in the world belonging to the family Papilionaceae. Generally the plants are erect and small shrubs. They grow usually to a height of two feet. The leaves are alternate, opposite, simple and stipulate. Small white or rose flowers are borne in clusters of few on short axillary racemes. The pods are splitting along both dorsal and ventral surfaces and are 11 to 12 ½ inches in length. Pods bear 8 to 10 seeds which vary considerably in shape, size and colour.

### Methods

The soil was cleaned absolutely by removing stones and other unwanted materials and was homogenized properly by mixing red soil and sand at 1 : 1 ratio and filled in pots having 7 kg capacity.

The study was conducted to assess the effect of the leavesNochi, Turmeric, Vasambu, Goose berry, Kuppaimeni) and biofertilizer (Rhizobium). In each pot different green manure (Nochi, Turmeric, Vasambu, Goose berry, Kuppaimeni) and biofertilizer was mixed individually. In to these treated pots ten cluster bean seeds were sown at two inches apart and one inch deep. Each treatment consisting of three replications was used for the experiment. The experiment was set up in completely randomized design. Pots were watered regularly. There were six treatments as given below which were evaluated against the control.

## Treatment details

$T_0$ - Control – Red loamy soil (7 kg)
$T_1$ - Red loamy soil + 35 g *Vitex nigundo* leaf powder / pot
$T_2$ - Red loamy soil + 35 g *Curcuma longa* rhizome powder / pot
$T_3$ - Red loamy soil + 35 g *Acorus calamus* rhizome powder / pot
$T_4$ - Red loamy soil + 35 g *Ribes uva-crispa* leaf powder / pot
$T_5$ - Red loamy soil + 35 g *Acalypha indica* leaf powder / pot
$T_6$ - Red loamy soil + 10 g Biofertilizer Rhizobium / pot

## Biochemical parameters

1. Chlorophyll – estimated on 30, 45, 60 and 75 DAS for leaves
2. Protein – estimated on 30, 45, 60 and 75 DAS for leaves
3. Total carbohydrates – estimated on 30, 45, 60 and 75 DAS for leaves
4. Ascorbic acid – estimated on 30, 45, 60 and 75 DAS for leaves.

## RESULTS AND DISCUSSION

The results of the experiments conducted on *Cyamopsis tetragonoloba* L. with the treatments of green leaf manure (Nochi, turmeric, vasambu, gooseberi and kuppaimeni) and biofertilizer (Rhizobium) on biochemical and parameters were statistically analysed and presented below.

## Chlorophyll Content

Chlorophyll 'a', 'b' and 'total' chlorophyll were higher in biofertilizer treatment on all the days. Chlorophyll 'a' content was higher in $T_6$ (biofertilizer) on 30 DAS (0.138 mg/g), 45 DAS (0.160 mg/g), 60 DAS (0.162 mg/g and 75 DAS (0.170 mg/g). The chlorophyll 'b' was higher in $T_6$ (biofertilizer) on 30 DAS (0.073 mg/g), 45 DAS (0.087 mg/g), 60 DAS (0.095 mg/g) and 75 DAS (0.106 mg / g). Total chlorophyll was found to be higher on biofertilizer ($T_6$) on 30, 45, 60 and 75 DAS (0.182 mg / g, 0.248 mg / g, 0.256 mg / g and 0.270 mg / g respectively). (Table I).

Chlorophyll 'a', 'b' and 'total' chlorophyll were minimum in control plants ($T_0$). Chlorophyll 'a' on 30, 45, 60 and 75 DAS (0.032 mg / g, 0.084 mg / g, 0.089 mg /g and 0.112 mg / g respectively), chlorophyll 'b' on 30, 45, 60 and 75 DAS (0.046 mg / g, 0.073 mg / g, 0.077 mg / g and 0.089 mg /g respectively) and total chlorophyll on 30, 45, 60 and 75 DAS (0.124 mg / g, 0.160 mg / g, 0.168 mg / g and 0.204 mg / g respectively).

Total chlorophyll content per plant at harvest increased significantly with increasing levels of rhizobium in cluster bean (Singh and Singh, 2005).

Biofertilizers affected chlorophyll, chlorophyll 'a' and 'b' significantly and the rhizobium culture produced significantly higher values of chlorophyll 'a' and 'b'. The beneficial effects of bacterial inoculation on increased chlorophyll content might have been due to the supply of higher amounts of nitrogen to the growing tissues and organs (Singh *et al.*, 1983).

**Table – IIII Influence of green leaf manure and biofertilizer on the chlorophyll contents of cluster beans (*Cyamopsis tetragonoloba* (L.))**

| Treatments | Chlorophyll 'a' (mg / g) | | | | Chlorophyll 'b' (mg / g) | | | | Total chlorophyll (mg / g) | | | |
|---|---|---|---|---|---|---|---|---|---|---|---|---|
| | 30 DAS | 45 DAS | 60 DAS | 75 DAS | 30 DAS | 45 DAS | 60 DAS | 75 DAS | 30 DAS | 45 DAS | 60 DAS | 75 DAS |
| $T_0$ – Control | 0.032 | 0.084 | 0.089 | 0.112 | 0.046 | 0.073 | 0.077 | 0.089 | 0.124 | 0.160 | 0.168 | 0.204 |
| **$T_1$ –** *Vitex nigundo(Nochi )* | 0.123 | 0.146 | 0.148 | 0.154 | 0.053 | 0.085 | 0.092 | 0.097 | 0.173 | 0.231 | 0.242 | 0.253 |
| $T_2$ – *Curcuma longa* (Turmeric) | 0.107 | 0.124 | 0.133 | 0.143 | 0.061 | 0.082 | 0.092 | 0.102 | 0.148 | 0.208 | 0.227 | 0.247 |
| $T_3$–*Acorus calamus* (Vasambu) | 0.041 | 0.085 | 0.153 | 0.160 | 0.065 | 0.075 | 0.094 | 0.104 | 0.139 | 0.161 | 0.249 | 0.265 |
| $T_4$-Acalypha indica (Kuppai meni) | 0.130 | 0.147 | 0.151 | 0.159 | 0.071 | 0.084 | 0.093 | 0.102 | 0.177 | 0.232 | 0.245 | 0.264 |
| $T_5$ - *Ribes uva-crispa* (Goose berry) | 0.118 | 0.120 | 0.129 | 0.140 | 0.069 | 0.083 | 0.092 | 0.100 | 0.179 | 0.204 | 0.223 | 0.243 |
| $T_6$ –Rhizobium | 0.138 | 0.160 | 0.162 | 0.170 | 0.073 | 0.087 | 0.095 | 0.106 | 0.182 | 0.248 | 0.256 | 0.270 |
| SEd | 0.00134 | | | | 0.00854 | | | | 0.00240 | | | |
| CD (0.05) | 0.00269 | | | | 0.01711 | | | | 0.00481 | | | |
| CD (0.01) | 0.00358 | | | | 0.02278 | | | | 0.00640 | | | |

SEd – Standard Error Deviation ; CD – Critical Difference ; DAS – Days After Sowing

The values indicate average of three replications.

## Protein

The maximum protein content was shown on 30$^{th}$ day by $T_3$ as (Vasambu rhizome powder) 31.47 mg / g and on 45 and 60 DAS it was higher in $T_5$ (Gooseberry) (33.80 mg / g and 35.86 mg /g). On 75 DAS it was higher in $T_6$ as biofertilizer (38.52 mg / g). The control plants registered lower content of protein 20.92 mg / g (30 DAS), 24.33 mg / g (45 DAS), 27.90 mg / g (60 DAS) and 29.51 mg / g (75 DAS) (Table – II ).

The higher proportion of biochemical constituents due to biofertilizer inoculation might be attributed to higher foliar nutrition, uptake of N, P, K and growth of seedlings. The above results are in corroboration with the findings of Jain and Trivedi (2005) who recorded that application of 19.65 kg P $ha^{-1}$ with rhizobium and PSB registered higher seed yield, oil yield and protein content in green gram.

Significantly more protein content was recorded in mulberry plants grown in the presence of green manure. According to Ravignanam and Gunathilagaraj (1996) the higher nutritional levels of mulberry is attributed to the increased root growth resulting in greater uptake of nutrients from soil. Sankar *et al.* (2000) observed significantly higher crude protein content in mulberry leaves by application of green manure.

## Carbohydrates

The highest carbohydrates was found in $T_6$ (biofertilizer) on 30 and 45 DAS (79.24 mg / g and 79.95 mg / g). On 60$^{th}$ day the highest carbohydrates content had been observed in $T_5$ (Gooseberry) 86.36 mg / g. On 75$^{th}$ day it was higher in $T_6$ (94.93 mg / g). The decreased content was in $T_0$, control on 30, 45, 60 and 75 DAS (47.74 mg / g, 62.73 mg / g, 78.10 mg / g and 85.41 mg / g). (Table – II ). Application of biofertilzier in combination with organic manure favoured higher accumulation of total carbohydrates in maize (Kamalakumari and Singaram, 1996). Increase in carbohydrate content by the application of biofertilizer may be due to availability of nutrients in higher amounts from the fertilizer.

**Table – II Influence of green leaf manure and biofertilizer on the protein, carbohydrates and ascorbic acid of cluster bean (*Cyamopsis tetragonoloba*(L.))**

| Treatments | Protein (mg / g) | | | | Carbohydrates (mg / g) | | | | Ascorbic Acid (mg / g) | | | |
|---|---|---|---|---|---|---|---|---|---|---|---|---|
| | 30 DAS | 45 DAS | 60 DAS | 75 DAS | 30 DAS | 45 DAS | 60 DAS | 75 DAS | 30 DAS | 45 DAS | 60 DAS | 75 DAS |
| $T_0$ – Control | 20.92 | 24.33 | 27.90 | 29.51 | 47.74 | 62.73 | 78.10 | 85.41 | 26.66 | 40.00 | 173.3 | 213.3 |
| $T_1$ – Vitex nigundo(Nochi ) | 26.28 | 29.92 | 31.28 | 31.90 | 57.41 | 67.06 | 80.72 | 87.95 | 38.33 | 53.33 | 186.66 | 220.00 |
| $T_2$ – *Curcuma long* (Turmeric) | 29.38 | 30.86 | 32.05 | 34.48 | 52.74 | 65.26 | 81.45 | 88.57 | 27.77 | 46.6 | 226.66 | 246.66 |
| $T_3$–*Acorus calamus* (Vasambu) | 31.47 | 32.48 | 33.44 | 34.60 | 51.29 | 68.22 | 81.53 | 86.91 | 33.33 | 46.6 | 220.00 | 246.6 |
| $T_4$-*Acalypha indica* (Kuppai meni) | 23.22 | 25.86 | 29.03 | 30.62 | 62.92 | 63.87 | 82.70 | 86.31 | 28.88 | 53.33 | 186.6 | 233.33 |
| $T_5$ - *Ribes uva-crispa* (Goose berry) | 25.99 | 33.80 | 35.86 | 37.03 | 66.12 | 74.03 | 86.36 | 90.90 | 33.33 | 60.00 | 206.66 | 246.66 |
| $T_6$ – Rhizobium | 28.61 | 31.34 | 35.78 | 38.52 | 79.24 | 79.95 | 87.47 | 94.93 | 40.00 | 80.00 | 246.66 | 266.66 |
| SEd | 3.33059 | | | | 5.92479 | | | | 17.27463 | | | |
| CD (0.05) | 6.67216 | | | | 11.86911 | | | | 34.60619 | | | |
| CD (0.01) | 8.88266 | | | | 15.80137 | | | | 46.07129 | | | |

SEd – Standard Error Deviation ; CD – Critical Difference ; DAS – Days After Sowing

The values indicate average of three replications

## ASCORBIC ACID CONTENT

Ascorbic acid content was analysed in leaves (Table – II). The highest ascorbic acid content had been observed in $T_6$ (biofertilizer) on 30 DAS (40.00 mg / g), 45 DAS (80.00 mg / g), 60 DAS (246.66 mg / g) and 75 DAS (266.66 mg / g). The lowest ascorbic acid content was observed in control ($T_0$) on 30 DAS (26.66 mg / g), 45 DAS (40.00 mg / g), 60 DAS (173.3 mg / g) and 75 DAS (213.33 mg / g). Biofertilizer application to maize resulted in an increased ascorbic acid, crude protein, starch, carbohydrates and phenol (Kamalakumari and Singaram, 1996). From this investigation it is concluded that the treatment containing biofertilizer (rhizobium), turmeric rhizome powder, *Acorus calamus* rhizome powder and *Acalypha* leaf powder could be an ideal and suitable potting mixture for better seedling and crop production in cluster bean (*Cyamopsis tetragonoloba* (L.) PNB).

## REFERENCES

Jain, P.C. and S.K. Trivedi, 2005. *Response of soyabean (Glycine max (L.) Merrill) to phosphorus and biofertilizers*. Legume Res., 28 (1) : 30-33.

Kamalakumari, K. and P. Singaran, 1996. *Quality parameters of maize as influenced by the application of fertilizers and manures.* Madras Agric. J., 83 (1) : 32-33.

Ravignanam, T. and K. Gunathilagaraj, 1996. *Effect of green manure on mulberry plant characters.* Madras Agric. J. 83 (6) *: 381-384.*

Sankar, M.A., M.C. Devaiah, Anitha Peter and B.T. Rangasamy, 2000. *Effect of graded levels of organic manure on growth, yield and quality of mulberry in relation to silkworm growth and cocoon production.* Crop Res., 19 (1) : 128-132.

Singh, A. and M. Singh, 2005. *Effect of farm yard manure, molybdenum and rhizobium on performance of cluster bean (Cyamopsis tetragonoloba L. Taub*), Indian J. Ecology 32 (2) : 109-111.

Singh, Jagdish Singh and Kalyan Singh, 1983. *Effect of phosphorus and biofertilizers on chlorophyll content of leaves and leghaemoglobin contents of fresh nodules in kharif grain legumes.* Indian J. Agron., 28 (3) : 229-234.